Developments in Clay Science and Construction Techniques

Edited by Amjad Almusaed, Asaad Almssad and Ibrahim Yitmen

Published in London, United Kingdom

Developments in Clay Science and Construction Techniques
http://dx.doi.org/10.5772/intechopen.1001628
Edited by Amjad Almusaed, Asaad Almssad and Ibrahim Yitmen

Contributors
Amjad Almusaed, Antonio Alberto Jaén Ortega, Asaad Almssad, Asmare Tezera, David Mubiru, Desalegn Abera, Heggadadevanakote Subbarao Prasanna, Ibrahim Yitmen, Luis Castillero, Maria De Los Angeles Ortega Del Rosario, Melany Medina, Nariman Hisham Halalo, Rafael Duque, Sam Bulolo, Tesfa Nega, Unnam Anil, Zenamarkos Bantie

First published in London, United Kingdom, 2024 by IntechOpen
IntechOpen is the global imprint of INTECHOPEN LIMITED, registered in England and Wales, registration number: 11086078, 167-169 Great Portland Street, London, W1W 5PF, United Kingdom

British Library Cataloguing-in-Publication Data
A catalogue record for this book is available from the British Library

Additional hard and PDF copies can be obtained from orders@intechopen.com

Developments in Clay Science and Construction Techniques
Edited by Amjad Almusaed, Asaad Almssad and Ibrahim Yitmen
p. cm.
Print ISBN 978-1-83769-607-9
Online ISBN 978-1-83769-606-2
eBook (PDF) ISBN 978-1-83769-608-6

For EU product safety concerns:
IN TECH d.o.o., Prolaz Marije Krucifikse Kozulić 3, 51000 Rijeka, Croatia,
info@intechopen.com or visit our website at intechopen.com.

Meet the editors

Professor Amjad Almusaed, Ph.D., a distinguished scholar in architecture focusing on environmental design, obtained his doctoral degree from Ion Mincu University, Romania. He further honed his sustainable and bioclimatic housing expertise with a postdoctoral research stint in 2004 at the School of Architecture, Aarhus University, Denmark. Prof. Almusaed's research primarily orbits around the ecological facets of architectural design and urban planning, contributing significantly to these fields. His academic journey is replete with extensive research endeavors, encompassing in-depth technical surveys and empirical studies. As a prominent figure in international architectural academia, Prof. Almusaed actively engages with various global architectural organizations, contributing profoundly to sustainable design and urban planning discourse. His editorial roles in numerous academic publications highlight his stature in the scholarly community. Prof. Almusaed's literary output is voluminous and impactful, with over 198 scholarly publications. These include articles, research papers, books, and chapters published in multiple languages. This extensive work underscores his relentless pursuit of advancing global knowledge in sustainable architecture and environmental design. His contributions are pivotal in shaping contemporary understanding and practices in these critical areas of study, solidifying his position as an eminent figure in environmental and sustainable architectural academia.

Associate Professor Asaad Almssad, a distinguished figure in construction and building sciences, has dedicated over 30 years to advancing industry, academia, and research. He has been integral to the academic faculty at prestigious institutions such as Umeå University and Karlstad University in Sweden. He has collaborated with various eminent European and international universities. His academic focus is deeply rooted in building materials and structural integrity, with a specialized interest in environmentally responsible construction principles and optimizing energy efficiency in building systems. This focus underscores his commitment to sustainable practices and innovation in construction engineering. With a prolific output of scholarly work, Associate Professor Almssad has contributed over 60 academic papers to revered journals, establishing himself as a leading expert in his field. His literary contributions also encompass a range of books that delve into the nuances of construction materials and sustainable building practices, reflecting his comprehensive understanding and expertise. Associate Professor Almssad holds a prestigious position as a docent at Karlstad University in Sweden. He continues imparting his extensive knowledge and experience to students and peers in this role, fostering academic excellence and innovation in construction science and engineering. His academic tenure is marked by his teaching and research contributions and his role in shaping the future o

Professor Ibrahim Yitmen, a distinguished scholar in architectural technology, obtained his Ph.D. from Istanbul Technical University, Turkey. Since February 2018, he has been a professor at Jönköping University, Sweden, in the Management of Construction Production department. His research integrates cutting-edge technologies in the built environment, focusing on digital twin–based smart built environments, augmented reality/mixed reality in cognitive buildings, and integrating digital twins with deep learning for advanced planning and construction. Dr. Yitmen is also deeply involved in exploring the application of blockchain technology in construction supply chains and the development of cyber-physical systems in the context of Construction 4.0. Dr. Yitmen has made notable contributions to academic literature, serving as the editor of influential books such as *BIM-Enabled Cognitive Computing for Smart Built Environment: Potential, Requirements, and Implementation* and *Cognitive Digital Twins for Smart Lifecycle Management of Built Environment and Infrastructure: Challenges, Opportunities and Practices*, both published by CRC Press. His editorial expertise was further recognized through his role as a guest editor for the MDPI journal *Applied Sciences*, specifically for the special issue on "Cognitive Buildings." Currently, Dr. Yitmen is leading several significant research projects in Sweden, funded by organizations like Smart Built Environment, Vinnova, and Jönköpings Läns Byggmästareförening. These projects underscore his commitment to advancing the integration of smart technologies in construction and built environment sectors, positioning him as a critical figure in sustainable and intelligent architecture.

Contents

Preface

In construction science, one of the most fundamental aspects has always been the interaction between conventional building materials and contemporary engineering. Clay, one of the oldest construction materials used by humans, continues to provide a fascinating combination of simplicity and complexity that has reverberated throughout history. Throughout its exploration of this dynamic topic, *Developments in Clay Science and Construction Techniques* sheds light on the lasting properties of clay and the cutting-edge developments that are changing its application in modern construction. The purpose of this book is to provide a compilation of thoughts from prominent scholars, engineers, and practitioners who have devoted their lives to gaining knowledge of clay and improving its qualities. To give a thorough, academically rigorous, and practically helpful overview, the chapters investigate various subjects, ranging from the microscopic behaviors of clay particles to large-scale architectural applications. The need to create durable and reliable structures has driven significant advancements in clay manipulation and use in recent years. Innovations in construction techniques enable more efficient and ecologically friendly building processes, while advancements in nanotechnology have opened new opportunities for improving the mechanical characteristics of clay-based materials. The innovations that are taking place are not only expanding the bounds of what can be accomplished with clay, but they are also contributing to the resolution of some of the most severe problems that are now being faced by the building industry. This book is written for a wide range of readers, including academic researchers, professionals working in industry, and students studying architecture, civil engineering, and materials science. This book provides a comprehensive grasp of both the fundamental concepts and the most recent breakthroughs in clay science and construction techniques. We hope that this book will not only provide a wealth of knowledge but also inspire further study and development, generating a new generation of building practices that are both technologically modern and profoundly anchored in tradition. We want to express our appreciation to every contributor who has contributed their knowledge and perspectives to this book. Their work is at the vanguard of a discipline that has been there for as long as civilization but is also undergoing constant change. We have high hopes that everyone interested in the future of building and the ongoing heritage of clay will find this book an invaluable resource and source of inspiration. The significant progress of construction methodologies and materials has always been fundamental to human progress. Out of all these materials, clay has maintained a distinct and long-lasting importance. Clay has demonstrated versatility and indispensability in various applications, from ancient adobe constructions to modern high-tech uses. The book *Developments in Clay Science and Construction Techniques* examines the many aspects of clay in contemporary construction. It aims to consolidate the most recent research, advancements, and practical uses of clay.

The book has two sections, each focusing on discrete yet interrelated facets of clay science and building. The initial segment, “Innovations and Applications of Clay in Construction”, explores the latest advancements and environmentally friendly techniques in utilizing clay materials. The second section, titled "Technical Studies and Effects on Clay Soil Properties", examines the scientific and technical research investigating the behavior and features of clay soils in different building scenarios.

Section 1: Innovations and Applications of Clay in Construction

Chapter 1: "Contemporary Innovations and Sustainable Practices in the Application of Clay Materials within Architectural Design and Construction Methodologies". The initial chapter establishes the context by exploring contemporary advancements and environmentally friendly methods linked to clay materials in architectural design and construction. Clay has become an essential resource in the building industry as it strives to find sustainable and eco-friendly materials. This chapter showcases the inventive methods architects and builders employ to integrate clay into modern constructions. The use of clay is being reinterpreted in the twenty-first century, including energy-efficient building envelopes and aesthetic upgrades. The chapter also explores how traditional techniques are modified to align with contemporary sustainability criteria. This demonstrates that traditional and modern approaches may live happily in the quest for environmentally friendly construction solutions.

Chapter 2: "Nanoclays as Fillers for Performance Enhancement in Building and Construction Industries: State of the Art and Future Trends". The second chapter explores the field of nanotechnology and its influence on the building sector, specifically focusing on the application of nanoclays. Nanoclays have demonstrated considerable potential as additives for improving the performance of construction materials. This chapter thoroughly examines the most advanced methods employed in integrating nanoclays and investigates potential developments that might transform the industry significantly. The subject encompasses enhancements in material robustness, longevity, and resilience against environmental elements. This chapter offers a prospective view of the possible impact of nanoclays on the advancement of construction materials by analyzing existing research and exploring future applications.

Chapter 3: "Advancing Sustainable Construction: Insights into Clay-Based Additive Manufacturing for Architecture, Engineering, and Construction". The third chapter presents the innovative notion of clay-based additive manufacturing, sometimes referred to as 3D printing, in architecture, engineering, and construction. This chapter examines how clay-based additive manufacturing enhances sustainability and enables unparalleled creativity and accuracy in architectural design and construction. The presented insights encompass technical improvements, case studies, and anticipated future developments. The building sector may attain heightened effectiveness and environmental friendliness by utilizing clay's inherent characteristics and the accuracy of additive manufacturing.

Section 2: Technical Studies and Effects on Clay Soil Properties

Chapter 4 examines the "The Effect of Samples Disturbance of Partially Saturated Expansive Clay Soils on the Soil Properties". In this chapter, the focus shifts toward a more technical analysis. It explores how sample disturbance affects the characteristics of partly saturated expansive clay soils. Studying the integrity of soil samples is crucial because it substantially impacts the accuracy of soil property evaluations, which in turn affects building decisions. This chapter presents comprehensive empirical data and analysis on the impact of disturbances on soil behavior, encompassing alterations in volume, moisture content, and structural integrity. Comprehending these impacts is essential for precise soil evaluation and formulating efficient building tactics in regions with expanding clay soils.

Chapter 5 is "Effect of Compaction Energy on the Behavior of Coefficient of Consolidation for the Compacted Fine-Grained Soils". In this chapter, the technical investigation focuses on the impact of compaction energy on the coefficient of consolidation for compacted fine-grained soils. Compaction is essential in soil preparation for construction, and the energy used can considerably impact soil characteristics. This chapter discusses the results of experiments that examine how different compaction energy levels affect the consolidation behavior of soils with a high proportion of small particles. The findings have significant ramifications for building methodologies, namely in enhancing soil stability and strength to guarantee the safety and longevity of structures.

Chapter 6 is titled, "Construction Techniques Related to Clay Soils: A Case Study in Africa". The last chapter presents a concrete case study from Africa, demonstrating the use of construction methods tailored to clay soils. This case study provides valuable insights into the difficulties and remedies in areas where clay soils predominate. By examining real-world examples, this chapter highlights the practical applications of the scientific and technical principles discussed in the previous chapters. The case study highlights the significance of comprehending the specific soil conditions in each area and adjusting construction methods accordingly to produce successful and environmentally friendly building projects.

Developments in Clay Science and Construction Techniques is a thorough investigation into the changing significance of clay in contemporary construction practices. This book comprehensively discusses how clay materials and soil features may be used to advance the building industry by combining innovative applications and technical studies. Every chapter enhances the existing knowledge repository by providing functional perspectives for scholars, practitioners, and policymakers. Given the industry's current focus on sustainability and efficiency, the insights and progress outlined in this book will undeniably significantly impact the future of construction.

This book consolidates and elucidates clay's various functions and consequences, drawing from the rigorous geology, ecology, and construction engineering disciplines. Clay's composition is shaped by sedimentary processes and geological origins, closely linked to mineral content. The properties of clay-rich soil are determined by these factors, resulting in the unique particle arrangement of clay, which in turn affects its many uses. Historical records emphasize the significant importance of clay in shaping construction techniques. Clay is widely used as a building material due to its easy availability, versatility, and ability to regulate temperature. This article comprehensively examines the evolution of clay-based building, comparing traditional methods with contemporary advancements. This text provides a comprehensive evaluation of the material characteristics of clay, its significance in architecture, and its environmental consequences, highlighting its crucial role in sustainable construction. Clay, a pliable and natural material, has played a vital role in sculpting, enabling artists to create complicated shapes and enduring works of art.

Amjad Almusaed, Ph.D.
Professor,
Department of Construction Engineering and Lighting Science,
Jönköping University,
Jönköping, Sweden

Asaad Almssad
Associate Professor,
Department of Engineering and Chemical Sciences,
Karlstad University,
Karlstad, Sweden

Ibrahim Yitmen
Professor,
Department of Construction Engineering and Lighting Science,
Karlstad University,
Karlstad, Sweden

Section 1

Innovations and Applications of Clay in Construction

Chapter 1

Contemporary Innovations and Sustainable Practices in the Application of Clay Materials within Architectural Design and Construction Methodologies

Amjad Almusaed, Ibrahim Yitmen and Asaad Almssad

Abstract

This chapter examines integrating innovative clay materials within modern architecture's environmental stewardship framework. Focusing on clay, it emphasizes its role in sustainable design and construction, driven by escalating ecological concerns and the need for green development. The chapter highlights clay's enduring appeal, resilience, energy efficiency, and eco-friendliness in architecture. It traces clay's historical use, from traditional bricks and terracotta to advanced composites, and examines significant advancements in production techniques that enhance material properties while reducing environmental impact. Sustainable clay extraction practices, lifecycle analysis, thermal efficiency, and clay's role in healthier indoor environments are discussed. Case studies illustrate contemporary architects' use of clay to meet esthetic, structural, and environmental needs, addressing barriers such as structural, economic, and regulatory challenges. Recommendations for modifying regulations, enhancing education, and embracing technological innovation are provided to promote clay's broader use in construction. The chapter concludes that clay should be significant in future architectural design and construction, driven by innovative and ecologically responsible approaches. It argues that strategic use of clay, combined with technology and ecological ethics, can achieve sustainable development goals and create environmentally responsible, efficient, and esthetically appealing built environments.

Keywords: sustainable architecture, clay material, green construction practices, environmental impact, manufacturing techniques

1. Introduction

Khan et al. argue that developing environmentally friendly composite materials for various uses arises from a growing consciousness about the environment and the imminent shortage of petroleum-derived resources [1]. This chapter provides a thorough analysis of the recent advancements and sustainable use of clay materials in the fields of architectural design and modern building methods. This chapter highlights the important role of clay materials in promoting environmentally friendly design and building techniques in response to increasing environmental concerns and the need for sustainable development. Abyzov et al. describe the crucial function performed by different material categories in tackling these challenges: recycled materials that provide opportunities for reuse; traditional, natural, and locally sourced building materials that, driven by technological progress, are now being reimagined for contemporary, environmentally aware, and energy-efficient structures (old-new natural materials); and nanomaterials combined with nanotechnologies [2]. The latter improves the quality and qualities of materials, thereby improving the environment and enabling the development of unique materials and architectural designs with exceptional properties. Clay materials are undergoing a significant reevaluation, not merely as traditional building blocks but as vital contributors to the advancement of sustainable construction practices.

The reevaluation of clay materials in construction reflects their attributes of energy efficiency, durability, and a minimal environmental footprint. According to Sev, the construction industry is critical in pursuing sustainable development by harmonizing economic growth with social progress and ecological conservation, employing a life cycle approach to sustainability [3]. This framework is founded on the principles of resource efficiency, lifecycle-oriented design covering the entire lifespan of construction projects, and the prioritization of human well-being alongside ecological health. A comprehensive historical analysis explores the usage of clay in construction, tracing its origins from ancient civilizations to its pivotal role in architectural evolution. Fernandes et al. argue that clay masonry represents one of the most ancient and enduring construction techniques, highlighted by its systematic evolution during the Mesopotamian, Egyptian, and Roman periods, with significant advancements in the Roman era enhancing the functionality of clay bricks [4]. This tradition has not only persisted but has been refined through the medieval period to the present, illustrating the material's adaptability and the unchanged simplicity that initially facilitated its adoption. Despite facing numerous environmental challenges across centuries, including extreme weather conditions and mechanical wear and tear, clay brick structures have demonstrated remarkable resilience, which is evident in their preservation and functionality in the 21st century. The discussion then progresses to a detailed categorization of contemporary clay-based materials, encompassing traditional bricks, terracotta, and synthesized clay composites. La Noce et al. highlight the increasing energy demands and consequent resource depletion, pointing out that interventions in the built environment are crucial for addressing these issues. Current evaluation methods primarily assess buildings based on their operational energy consumption and emissions, underscoring the need for comprehensive approaches encompassing construction materials' environmental impacts throughout their lifecycle [5]. However, a building operates as an open system, involved in an ongoing flow of energy, materials, and social and economic connections with its environment, extending beyond its period of use. Therefore, improving a building's energy efficiency requires a comprehensive analysis of these interactions across the project's entire lifespan, organizing this communication to promote a more environmentally friendly architectural future.

In contemporary architectural practices, it is imperative to comprehend the nuanced characteristics of diverse materials and their critical roles in fulfilling the esthetic, structural, and ecological demands of modern architecture. This analysis focuses on recent technological advancements that have significantly transformed the production and application of clay-based materials. Advanced manufacturing techniques have been developed to enhance the physical properties of these materials while concurrently reducing their environmental impact. Among these innovations, Barry Berman's research underscores the adoption of 3D printing technology in the construction industry. This method enables the cost-effective production of customized items in small batches. Further scholarly discussion suggests that 3D printing may substantially reduce the logistical advantages traditionally associated with manufacturing in economies with lower labor costs, primarily by minimizing the labor force required in production facilities.

This analysis showcases the evolving landscape of material science in architecture. It posits a future where sustainable and personalized construction methods become increasingly prevalent, reshaping our approach to architectural design and its environmental impact [6]. This section examines the synergistic relationship between traditional material science and contemporary sustainable construction practices. It delves into environmentally sustainable methodologies for utilizing clay, emphasizing lifecycle analysis, thermal efficiency, and the contributions of clay materials to enhancing indoor environmental health. In this context, Hussain, Anwar, and Mohammad Arif Kamal posit that within the dynamic landscape of the construction industry, planners, architects, engineers, and builders are pursuing innovative materials and technologies for future building projects. These innovations deliver many benefits, including enhanced energy efficiency, conservation of resources and water, improved indoor air quality, reduced life cycle costs, and augmented durability [7].

This discourse emphasizes the essential integration of sustainable materials such as clay into architectural designs, addressing the ecological challenges of modern construction while enhancing occupant health and well-being. Clay is portrayed as crucial in evolving construction techniques that align with global sustainability goals. The narrative includes case studies that illustrate intentional uses of clay in cutting-edge architectural designs, notably Saeed Sakhdari's innovative approach that combines 3D printing technology with clay, utilizing post-tensioning techniques to create structurally reinforced clay artifacts. This novel method synergistically blends the natural flexibility of clay with the enhanced stability provided by post-tensioning, showcasing a pioneering approach to construction methods [8]. The chapter aims to construct a pavilion visually demonstrating the system's effectiveness in practical scenarios, highlighting how innovative solutions leverage clay's distinctive properties to tackle esthetic, structural, and ecological challenges.

Moreover, the text explores the barriers to widespread clay usage in the construction industry, including structural, financial, and regulatory challenges. It advocates for a comprehensive strategy that encompasses enacting new regulations, initiating educational campaigns, and encouraging technological advancements to address these issues. This approach is designed to hasten the broader integration of clay materials into sustainable construction practices and underscores the necessity for collaborative efforts among policymakers, educators, industry stakeholders, and researchers. In conclusion, the chapter calls for a reevaluation of the role of clay materials in architectural design and construction, advocating for innovative and environmentally conscious methodologies. This theory proposes that by strategically utilizing clay, technical advancements, and ecological principles, we may significantly contribute to attaining sustainable development objectives. This vision advocates for a paradigm shift in the way we design eco-friendly and efficient spaces that are also esthetically pleasing. It represents a new age in sustainable design and construction that adheres to ecological stewardship and technology innovation concepts.

2. Clay in sustainable architecture: bridging innovation and global goals

2.1 Foundational strategies for eco-friendly building practices using clay

The use of clay in environmentally responsible building practices, which has been around for millennia, is gaining renewed interest in contemporary architecture due to its numerous advantages regarding the environment, the economy, and society [9]. This approach is rooted in principles prioritizing using locally sourced resources, energy efficiency, and long-term durability throughout construction. Clay, a naturally abundant material, is recognized for its excellent thermal mass properties, contributing to substantial energy savings in construction by maintaining consistent interior temperatures [10]. Furthermore, clay materials are renowned for regulating indoor humidity levels, enhancing air quality and occupant comfort. Throughout its lifecycle in the construction industry, from extraction and processing to disposal or recycling, clay has a minimal environmental impact [11]. Some examples of sustainable practices include minimizing waste during the manufacturing process and increasing the ability of a structure to be recycled once it has reached the end of its useful life. In architecture, including clay materials often reflects a deep respect for local environments, cultures, and traditional construction techniques [12]. Utilizing clay promotes ecological sustainability and supports the continuity of social and cultural practices. Economically, sustainable clay construction aims to reduce operational costs associated with heating and cooling, bolster local businesses, and create durable buildings that necessitate minimal maintenance over their lifespan [13]. This holistic approach to building with clay highlights its crucial role in developing resilient communities equipped to address the challenges posed by climate change and resource scarcity. Employing clay in sustainable building practices exemplifies a comprehensive strategy aligned with global sustainability objectives [14]. Using clay's inherent benefits to generate environmentally friendly, economically feasible, and socially helpful structures is crucial to sustainable design and construction because it uses clay's already-existing qualities.

2.2 The exploring architectural innovations and practices for sustainable development goals: integrating clay as a sustainable material

The relationship between the Sustainable Evolution Goals (SDGs) and architecture highlights a critical connection in the evolution of the built environment. This connection underscores the pivotal role that architectural practices play in meeting global sustainability standards [15, 16]. Introduced by the United Nations in 2015, the Sustainable Development Goals (SDGs) succeed the Millennium Development Goals (MDGs), aiming to address their unmet targets by expanding their agenda. The SDGs are designed to eradicate poverty, protect the environment, and foster global peace and prosperity by 2030, establishing a broad framework for the well-being of humanity and the planet. This framework emphasizes the urgent global issue of severe poverty and promotes worldwide peace and freedom [17]. Within this strategic framework, architecture directly contributes to specific objectives, such as promoting sustainable cities and communities (Goal 11), supporting responsible consumption and production (Goal 12), and facilitating collective action against climate change (Goal 13). The Sustainable Development Goals, in conjunction with the New Urban Agenda, recognize the crucial significance of urban settings in promoting sustainable development. However, national governments officially approved these frameworks, which required modification or localization to execute these agendas successfully at the municipal or local level. This highlights the importance of developing customized approaches that synchronize national goals with specific local conditions, enabling the implementation of these worldwide aims in various urban settings [18]. Architecture significantly contributes to sustainable development by engaging in sustainable urban planning, constructing energy-efficient buildings, and utilizing eco-friendly materials like clay. Recognized as a crucial element for environmental sustainability, green construction leverages locally sourced, eco-friendly building materials that adhere to eco-design principles. This approach fosters habitats that promote healthy living and respect the cultural and architectural heritage, which is vital in conserving natural resources and aligning with broader sustainable development goals [19].

Moreover, sustainable design aims to reduce the environmental impact of built environments by optimizing materials, energy, and spatial usage. This strategy extends to the environmentally conscious renovation and adaptive reuse of existing structures to meet stringent ecological standards. Clay is particularly notable in sustainable design for its abundant availability, energy-efficient production processes, and superior thermal properties, making it a model material for sustainability in architectural practices. The utilization of clay in architectural projects, including conventional brickwork and novel clay composites, supports the Sustainable Development Goals (SDGs) by improving the effectiveness of construction, minimizing waste, and harnessing renewable resources. On the other hand, the urgent worldwide problem of environmental deterioration and the limited availability of resources has caused a change toward sustainable and circular economic processes in many industries.

The roofing and composite sectors are integral to the building and infrastructure industries, significantly contributing to the shift toward sustainability. This trend demonstrates a broader commitment to sustainable practices, transforming construction techniques by adopting materials and methods that reduce environmental impact and conserve resources [20]. In line with Goal 11, which advocates for cities to be sustainable, inclusive, resilient, and safe, using clay materials allows architects and urban planners to create spaces that enhance accessibility, emphasize environmental stewardship, and foster community engagement, ultimately improving urban life quality. Through the incorporation of green building methods, the improvement of walkable surroundings, and the facilitation of efficient public transportation networks, the adaptability of clay enables sustainable urbanization. Goal 12 advocates for responsible consumption and production, corresponding to using clay in construction. This material choice exemplifies a commitment to sustainability, reducing environmental impacts using recyclable and durable materials [21]. Employing clay in construction not only diminishes ecological degradation but also supports principles of the circular economy by facilitating extensive material reuse and recycling opportunities. Consistent with Goal 13, which focuses on climate action, structures incorporating clay help mitigate the effects of climate change. Due to its high thermal mass, clay is used to construct buildings that maintain stable temperatures, naturally cooling in the summer and retaining warmth in the winter. This attribute decreases the dependency on mechanical heating, ventilation, and air conditioning (HVAC) systems, thus lowering emissions of greenhouse gases.

One way to lessen the environmental toll of building materials is to adopt sustainable methods for extracting and treating clay. Incorporating clay into building procedures by the SDGs framework reveals a well-thought-out approach to achieving social, economic, and environmental sustainability. Using a creative approach that places clay at the center of its construction, architecture guides society toward a more equitable and sustainable future, ensuring that our world is managed responsibly for future generations [22].

3. Figurative synthesis of sustainable design elements: interconnecting chromatic strategies, spatial quality, and biophilic principles within clay architecture

3.1 Clay integration in biophilic design: a multidisciplinary approach for sustainable architecture

Clay, a material deeply ingrained in the annals of human history, has experienced a renaissance within biophilic architecture. Biophilic design, a topic in architectural discussions, advocates for the restoration of a deep and inherent bond between humans and the natural world. This is accomplished by intentionally combining natural materials, maximizing natural light, including greenery, and using other features that evoke the outdoors [23]. The technique is grounded in "Biophilia," a concept that describes humans' intrinsic emotional connection with biological systems influenced by evolutionary learning processes. This affinity prompts diverse emotional responses, such as attraction, rejection, serenity, and anxiety, often induced by fear. The increasing incorporation of biophilic principles into design practices has led to a deeper understanding of their capacity to substantially enhance physical and mental well-being, diminish stress, and stimulate creativity. Despite the increasing interest in this topic, scholars still need to clarify the psychological effects of real vs. fake natural elements in indoor settings on people's health and well-being [24, 25]. The incorporation of clay in biophilic architecture is highly esteemed. Its resurgence in contemporary construction represents a nod to tradition and a strategic choice aligned with the core principles of sustainability, wellness, and ecological consciousness inherent to biophilic design. This reassessment of clay represents a broader recognition of biophilic elements' potential to enhance physical and mental health, satisfaction, and productivity in communal workspaces and educational settings. A research methodology was developed to empirically assess these claims, based on the Flourish Model. This approach involved collecting qualitative data through interviews, quantitative indoor environmental quality (IEQ) measurements, and post-occupancy evaluation (POE) feedback. This methodological framework aimed to comprehend how incorporating biophilic design, mainly through clay, thoroughly influences the interior environment and, therefore, the experiences and results of those residing in these areas [26]. The integration of clay within biophilic architecture extends beyond mere esthetic appeal, embodying a holistic approach to sustainable design. Its significant thermal mass is pivotal in passive climate regulation, absorbing and dissipating heat to naturally balance indoor temperatures and minimize reliance on artificial heating and cooling systems. Over the past decade, the concepts of 'nature' and biophilic design have garnered increasing interest in the architectural community, particularly in response to escalating environmental concerns. Despite this growing focus, there remains to be more debate and uncertainty over how 'nature' is conceptualized and implemented in both practical and research contexts [27]. Additionally, contributing to the construction of comfortable living and working settings that are in tune with natural cycles, this skill improves energy efficiency and creates such surroundings. Clay's hygroscopic properties facilitate the regulation of indoor humidity, as it absorbs surplus moisture in humid environments and releases it under drier conditions.

DOI: http://dx.doi.org/10.5772/intechopen.1005787

This capability is crucial for mitigating mold growth and the proliferation of airborne pathogens closely associated with respiratory health challenges. **Figure 1** conceptualizes the interrelationships between various aspects of architectural design, focusing on clay architecture. Central to the diagram is 'Clay Architecture,' positioned as the foundation from which the other concepts radiate, emphasizing its pivotal role in sustainable and innovative building design.

Clay's inherent capacity to regulate indoor air quality is vital, as it reduces mold proliferation and airborne disease transmission. Additionally, the psychological benefits conferred by clay within the biophilic design are considerable. The material's tactile properties and organic colors establish a strong connection with the natural environment, fostering calmness and enhancing mental well-being among users. Research findings underscore substantial evidence supporting certain aspects of biophilic design, notably including natural elements. Nevertheless, further empirical research is necessary to substantiate other dimensions, such as the employment of natural materials or processes [28]. In biophilic design, this psychological comfort is of utmost importance. This approach to design considers the psychological and cognitive connections associated with architectural spaces to be just as important as the physical and functional qualities of the rooms.

Clay is a versatile building material that offers architects and designers significant technical and artistic latitude. It can be utilized across a wide range of construction applications [29]. Clay is a versatile material that can be transformed into multiple structural and esthetic forms. This includes the production of bricks that can support weight, walls that can store heat, and advanced tiles and facades, all of which contribute to improving the presence of natural elements in architecture. Clay is employed in various capacities, including constructing durable structures, fabricating standard building walls, decorating interior and exterior surfaces, producing paving materials, and integrating them into contemporary artworks. This extensive array of applications

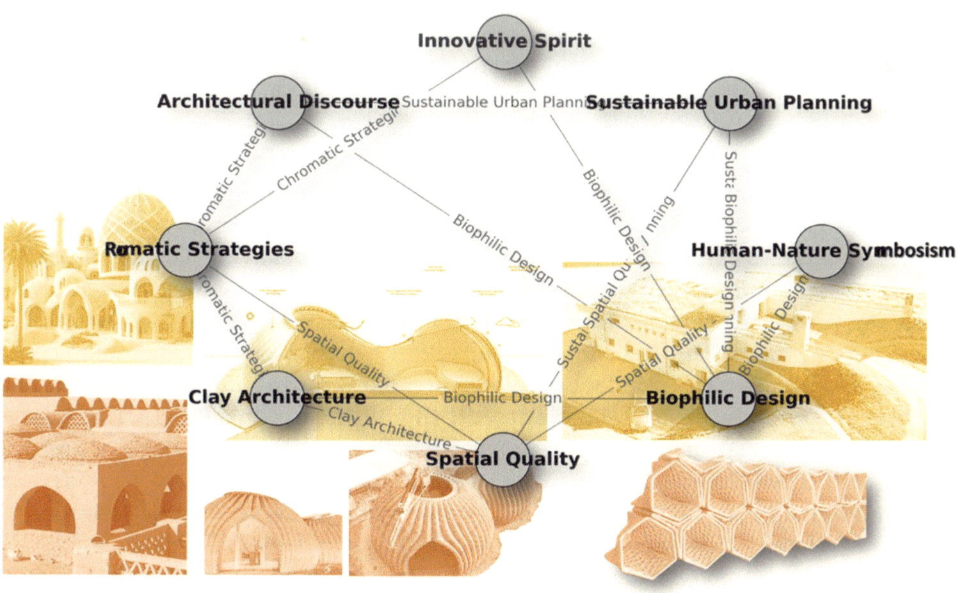

Figure 1.
Synergistic approaches to clay architecture and biophilic design within sustainable urban development.

underscores clay's remarkable versatility in facilitating diverse architectural expressions [30]. Adobe, considered one of the oldest bricks in the Western Hemisphere, explains the lasting significance of clay in the development of construction materials and architectural techniques. This innate adaptability enables the creation of structures that are responsive to the environment and aligned with cultural and contextual factors, reflecting concepts of biophilic design. It facilitates a sophisticated connection between the constructed surroundings and natural and socio-cultural ecosystems. Academic conversations emphasize the revival of clay in architecture that promotes a connection with nature while aligning with sustainable development goals and the building industry's need to reduce environmental impact. Clay is a cornerstone in biophilic design. It is valued for its abundant availability, potential for local production, and recyclability, harmonizing with the principles of a circular economy and low-impact architecture. Its use in construction markedly diminishes the environmental footprint of buildings, fostering biodiversity preservation and natural resource conservation, thereby aligning with overarching sustainability objectives. The incorporation of clay in biophilic architecture embodies a comprehensive building approach that deeply recognizes and honors the innate human inclination for connection with the natural surroundings. This method addresses the environmental and health challenges of sustainable living while enriching architectural design's esthetic and cultural dimensions. Clay exemplifies the integration of traditional materials with contemporary design principles to forge sustainable, nature-connected living spaces. This approach in architectural design advances beyond mere functionality, emphasizing deep respect for and integration with the natural environment, promoting a model of sustainability and meaningful engagement with nature.

3.2 Clay chromatic strategies and biophilic principles in clay architecture: an integrative approach to sustainable building design

The use of clay and color in contemporary architectural practices extends beyond mere esthetic considerations, necessitating a detailed exploration of their impact on built environments' functionality and emotional resonance. The subjective experience of perceiving different hues and color combinations is also recognized. Jaglarz highlights the crucial significance of color as a critical element in the arrangement of architectural spaces, exerting a substantial impact on the observer's perception [31]. This acknowledgment underscores the significance of color in architectural design, mainly due to the challenges in predicting color esthetics. It highlights color's profound influence on human perception, cognition, emotions, and behavior. The literature evaluations and survey data provide valuable insights and conclusions as a foundation for defining different color strategies in architectural design. The discussion is enhanced by including the ideas of biophilic design, which promotes using natural materials and processes in architectural structures to cultivate a mutually beneficial interaction between humans and their natural environment. Amjad and Asaad classify their work into three primary categories: Nature in Space, Natural Analogues, and Nature of Space, while introducing an initial set of 'biophilic conditions.' The scientific foundation of biophilic design has since been expanded and deepened through advances in related fields such as neurology and endocrinology [32]. An in-depth examination is required to understand how combining chromatic schemes and biophilic design principles in clay architecture affects clay structures' spatial quality and environmental sustainability. Here, it is necessary to investigate the complex interaction between architects, artists, and manufacturers and the fundamental principles of biophilia, particularly within the context of Ivrea. Amjad and Asaad's research clarifies that biophilic design goes beyond simply incorporating plants into architectural structures or choosing pleasing color combinations. It dramatically expands the range of architectural variety by including a wide range of natural components that involve the physical, sensory, symbolic, morphological, material, and spiritual dimensions of human experience [33]. This discussion highlights the importance of using color, clay materials, and natural features in architectural design. This highlights how such a comprehensive approach enhances buildings' physical and visual aspects and imbues urban environments with more liveliness and a shared identity. The combination of these elements has been shown to significantly impact the overall design and character of places, resulting in a more unified and inviting atmosphere in urban areas. Moreover, this discourse expands to include the significant influence of esthetic beauty on the overall welfare of society, placing it as a crucial factor for promoting exploration, acquisition of information, education, and curiosity. The deliberate choice of materials that align with the functional nature of buildings is believed to significantly enhance structures' esthetic appeal and long-term viability, thus contributing to their overall beauty and sustainability [34]. These buildings are portrayed as essential elements of the city, representing the essence of civilization and highlighting the direct connection between establishing cultural and educational spaces and the sustainable growth of metropolitan areas. This perspective underscores the critical importance of these structures in improving urban environments, especially on city outskirts where development disparities frequently occur. The Olivetti era was marked by an inventive approach characterized by the strategic use of color combined with biophilic elements [35]. **Figure 2** represents a conceptual framework depicting the interrelations between

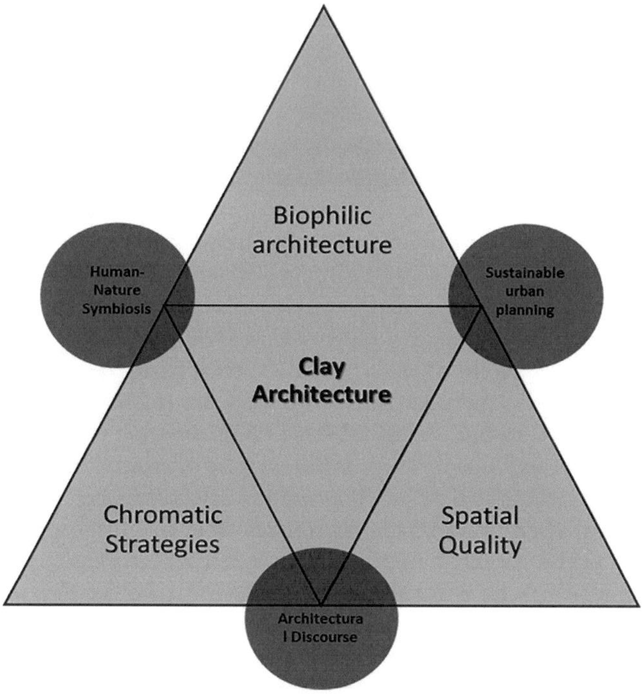

Figure 2.
Triadic framework of clay architecture: synthesizing chromatic strategies, spatial quality, and biophilic principles.

various architectural design elements centered around clay architecture. The triangle represents a stable structure with clay architecture as the foundational base, suggesting its central role in architectural practice.

This ethos, characterized by the seamless integration of esthetic appeal, technological innovation, and environmental consciousness, promotes the development of community identities through the meticulous application of these principles [36]. The transformational potential of color and biophilic design in architecture is brought to light by this research, which advocates for the purposeful incorporation of these aspects as essential components in sustainable urban planning and design approaches. The chapter highlights the crucial relevance of biophilic principles in decreasing the ecological footprint of buildings, boosting biodiversity, and improving human well-being. These goals may be accomplished by emulating the richness and diversity of natural habitats inside architectural structures. As a result, this inquiry substantially contributes to the architecture debate by creating a paradigm that promotes sustainability, health, and esthetic richness in the built environment. This paradigm aligns with the overarching aims of resilient and harmonious urban life [37].

4. From earth to innovation: analyzing clay's evolution in architecture from traditional methods to smart adaptive solutions

4.1 Innovative applications of clay in modern architecture: case studies and sustainability insights

An enhanced understanding of the innovative application of traditional materials in contemporary architectural practices can be achieved by examining modern clay architecture *via* case studies. This approach underscores the material's diverse utility in fostering sustainability, esthetic appeal, and environmental integration. Additionally, exploring new alternatives and technological solutions in designing contemporary individual houses using clay architecture is essential. Clay, recognized as a natural and eco-friendly building material, is experiencing a resurgence in modern projects due to its ecological and esthetic advantages, garnering the attention of an increasing number of architects. Contemporary research efforts are concentrated on augmenting its durability and strength, aiming to render clay a competitive and viable material for contemporary architectural endeavors [38]. Clay's flexibility in current building technologies is brought to light by this research, emphasizing the significant contributions that clay has made to ecological design and sustainable urban development. In the School of Art and Design setting in Saint-Étienne, France, which LIN Architects built, terracotta cladding appears to blend esthetic refinement with practical functionality and sustainability. The concept leverages terracotta's insulating properties to enhance energy efficiency by minimizing reliance on artificial climate control systems and offering a visually appealing exterior. This chapter demonstrates the twin benefits that clay materials offer in contemporary architecture: their contribution to the visual character of the structure as well as their function in the conservation of energy within the building. Similarly, the Earth Architecture Project, which the Nka Foundation in Ghana is carrying out, is an example of how local clay resources may be utilized to promote environmentally responsible and culturally significant methodologies for construction. The Nkabom House presents an innovative approach to sustainable building in Ghana, utilizing earth construction and recycling methods. As depicted in **Figure 3**, this case study exemplifies using

Figure 3.
The Nkabom House: a model for sustainable earth construction and recycling in Ghana.

local materials and traditional techniques to create a cost-effective, environmentally friendly structure aligned with the region's cultural context and climatic conditions.

Because of this endeavor, the material's diversity is highlighted in its ability to provide cost-effective, environmentally harmonious, and culturally relevant architectural solutions. This project presents a scalable and sustainable architectural development paradigm that integrates contemporary design principles with traditional mud-brick techniques, applicable in politically and economically similar regions. Additionally, Potter's House in Australia, designed by Peter Southbury Architecture, exemplifies the strategic use of clay bricks and tiles for thermal efficiency and esthetic integration with the landscape. This approach highlights the thermal mass advantages of clay, facilitating natural temperature regulation within the building [39]. The project highlights the symbiotic link between contemporary clay architecture and the natural setting in which it is situated, highlighting the material's role in addressing issues of sustainability and place-making. When taken as a whole, these case studies contribute to a more in-depth comprehension of the capability of contemporary clay architecture to meet contemporary concerns relating to sustainability, environmental stewardship, and cultural continuity. These projects underscore the enduring relevance of traditional materials within contemporary architectural discourse by redefining the application of clay in building structures. Additionally, they provide valuable insights into integrating sustainability principles within the architectural design of the built environment [40]. **Table 1** provides a comprehensive overview of the project name, the architect or organization responsible, the significant characteristics of the project in terms of clay utilization, and valuable insights into the sustainability contributions of each project.

Project	Architect/organization	Key features	Sustainability insights
School of Art and Design, Saint-Étienne	LIN Architects	Terracotta cladding, insulating capabilities, energy efficiency, esthetics, and sustainability.	Reduces dependency on artificial climate control and contributes to sustainable urban development.
Earth Architecture Project, Ghana	Nka Foundation	Local clay resources are environmentally responsible, culturally significant, and cost-effective.	Promotes environmentally responsible construction and scalable and sustainable architectural development.
Potter's House, Australia	Peter Southbury Architecture	Clay bricks and tiles, thermal efficiency, esthetic harmony with terrain, natural temperature regulation.	It highlights the thermal mass benefits of clay and addresses sustainability and place-making.

Table 1.
Sustainable architectural innovations: a study of terracotta, earth, and clay constructions.

4.2 A comparative study of traditional and modern architectural techniques with a focus on clay materials

The transition from traditional to contemporary architectural practices presents a rich domain for in-depth comparative scrutiny, illuminating the profound influences of technological progress, advancements in materials science, and heightened environmental awareness on the architecture and construction industries. Despite the construction sector's inherent conservatism and risk aversion characteristics, this analysis concentrates on incremental enhancements attainable throughout the value chain. Notably, the adoption of supplementary cementitious materials and the optimization of cement's clinker content are emphasized. Crucially, it is posited that the aggregate impact of these incremental improvements could significantly contribute to mitigating greenhouse gas emissions by as much as 50%, contingent upon comprehensive stakeholder engagement [41]. Consequently, it is essential to highlight the varied methodologies and materials used across different eras, reflecting the evolving objectives and challenges architects and builders face. Traditional architectural practices, deeply rooted in local cultures, climates, and readily available materials like stone, wood, and clay, inherently address environmental and societal demands with a focus on sustainability. For example, using clay exemplifies this commitment to eco-friendliness, with its superior thermal mass properties facilitating natural heating and cooling, thereby emphasizing the sustainability inherent in traditional techniques. These approaches also incorporate passive solar design, natural ventilation, and other environmentally conscious strategies aligned with contemporary sustainable design principles. The pressing concern of climate change and the imperative to mitigate CO_2 emissions have significantly influenced the selection of outdoor plant species. Consequently, evaluating CO_2's environmental effects on plants has become integral to the decision-making process. Significantly, mitigating greenhouse gas (GHG) emissions from buildings is crucial in addressing the impacts of climate change and energy shortages [42]. However, while traditional methods excel in environmental compatibility and cultural significance, they often must catch up to modern practices' precision, durability, and scalability. Conversely, modern architectural techniques are characterized by the innovative use of concrete, glass, steel, and advanced versions of clay materials. These contemporary practices enable greater design flexibility, allowing for the construction of taller and more complex structures. Introducing engineered clay products, such as intelligent clay bricks that can adapt their properties to environmental changes, exemplifies the integration of traditional materials with cutting-edge technology [43]. Moreover, computational design tools enhance energy efficiency and structural integrity, addressing today's sustainability and climate resilience demands. However, the reliance on industrial materials and processes can sometimes lead to increased carbon footprints and a detachment from local context and heritage. **Table 2** offers a comparative analysis of traditional versus modern architectural techniques, emphasizing the evolution of materials and methodologies, focusing on clay materials.

The comparative analysis of traditional versus modern techniques, especially considering the role of clay materials, reveals a continuum of trade-offs between cultural continuity, environmental sustainability, and technological innovation [44]. While traditional methods offer deep connections to place and environment, modern techniques solve the challenges of urbanization and climate change. Bridging these diverse approaches by leveraging the strengths of each suggests a pathway toward architectural practices that are more sustainable, resilient, and culturally enriched. This synthesis not only narrows the gap between past and present but also clears the way for future innovations that respect heritage while embracing progress, with clay materials serving as a testament to the enduring relevance of natural resources in the built environment [45].

Aspect	Traditional architectural techniques	Modern architectural techniques
Core Materials	Stone, wood, clay	Concrete, glass, steel, advanced clay products
Methodology	Direct response to environmental and societal needs, sustainability-focused	Innovative use of materials for flexibility, construction of complex structures
Sustainability	High, with practices like passive solar design, natural ventilation	Variable aims for energy efficiency and climate resilience but higher carbon footprints
Technological Integration	Low focuses on natural and locally available materials	High, employs computational design tools, engineered materials
Environmental Impact	Low, utilizes eco-friendly materials and methods	Potentially high due to industrial processes and materials
Cultural Significance	Deeply ingrained in local cultures and traditions	Sometimes detached from local context and heritage
Scalability and Durability	Limited by materials and techniques	Enhanced by technological advancements
Flexibility in Design	Constrained by materials and traditional methods	High, allowing for taller and more intricate designs
Innovation	Emphasizes sustainability and traditional knowledge	Integrates cutting-edge technology with traditional materials like clay
Goal	To achieve sustainability and fulfill societal needs within environmental constraints	To address urbanization and climate change challenges, enhancing structural integrity

Table 2.
Comparative analysis of traditional and modern architectural techniques: emphasizing the evolution and application of clay materials.

4.3 Advancements in architectural design: the role of smart clay materials in developing adaptive building solutions

Integrating intelligent clay materials into adaptable building solutions marks a significant paradigm shift, blending the enduring qualities of traditional clay-based construction with the transformative potential of contemporary technological innovations. This synergistic approach has catalyzed the development of a new generation of architectural methodologies, which exploit clay's intrinsic thermal properties and environmental sustainability, augmented by advanced technological systems [46]. By facilitating the construction of buildings that are not only environmentally conscious but also dynamically sensitive to changing climatic circumstances, these technologies make it possible to create structures that maximize the comfort of their occupants while reducing the amount of energy used. One of the most critical aspects of developing innovative clay materials is the introduction of embedded sensors and actuators into traditional clay-building elements like bricks and tiles [47]. The materials can autonomously monitor ambient environmental conditions, including temperature, humidity, and light levels, allowing them to adjust their physical properties in real-time. For example, the modulation of porosity in bright clay bricks, which makes natural ventilation easier, is an example of a mechanism through which these materials actively regulate the temperature inside the building. This reduces the reliance on artificial heating and cooling systems, which in turn increases the ecological footprint of the building. In addition, the adaptive capabilities of these materials go beyond environmental management and include the structure's robustness. Intelligent clay materials bolster structural resilience against natural disasters, enhancing the built environment's safety and longevity. Recent advancements in clay composites, engineered to mitigate seismic stresses, underscore their significant potential. The research and development of these intelligent clay materials, integrated with adaptable building solutions, represents an innovative synthesis of traditional construction techniques and contemporary technological innovations. This interdisciplinary approach challenges established architectural paradigms and heralds a future where buildings transcend their static functions, evolving into dynamic, sustainable habitats that actively contribute to the well-being of their occupants and the broader environmental context. This evolution highlights a captivating narrative of development within sustainable design and the crucial role of new material science in defining the future of built environments.

5. Integrating sustainability into clay construction: innovations, strategies, and environmental impact

5.1 Exploring innovations in clay material research: bridging disciplinary frontiers

New research trends in clay materials are transforming their applications and our understanding across multiple scientific fields. This discussion emphasizes clay's role in advancing sustainable development, environmental engineering, nanotechnology, and biological sciences, showcasing its cross-disciplinary importance and leading-edge contributions. Clay-based materials are rapidly included in sustainable building approaches because of their inherent thermal mass, durability, and eco-friendly characteristics. Innovative research efforts are being directed at enhancing these features

Field	Impact of innovations in clay material research
Sustainable Architectural Practices	Enhancement of thermal mass, durability, and ecological sustainability in construction methodologies.
Ecological Engineering	Developing sophisticated filtration systems for water and soil remediation; efficient sequestration of heavy metals and organic pollutants.
Advanced Material Science	Synthesis and integration of clay nanoparticles within polymer matrices, yielding composites with superior mechanical, thermal, and electrical properties.
Biomedical Applications	Pioneering contributions to tissue engineering, regenerative medicine, and targeted drug delivery mechanisms, leveraging biocompatibility and modifiable surface chemistry.

Table 3.
Impact of clay material research across disciplines.

to mitigate the environmental repercussions of building operations [48]. Efforts are underway to replace conventional construction materials with energy-efficient, structurally sound clay composites. In ecological engineering, clay minerals' adsorption properties are used to develop advanced filtering systems for cleaning water and soil, effectively sequestering heavy metals and pollutants. This offers a sustainable solution to environmental pollution. Meanwhile, combining clay nanoparticles into polymers and composites in nanotechnology has produced materials with exceptional mechanical, thermal, and electrical properties. Under their increased performance metrics, such as lightweight characteristics, robustness, and better barrier functions, these nanoclay composites significantly accelerate advancements in various industries, including the automotive and electronics industries [49]. In addition, clay materials are developing many applications in biomedicine, notably in tissue engineering, regenerative medicine, and drug delivery systems. **Table 3** illustrates how clay material research advances sustainability, environmental health, technology, and medicine. Highlighting clay's role in innovative solutions underscores its significance across multiple disciplines, addressing global challenges.

Clay particles help construct precise drug delivery systems and scaffolds that stimulate cellular regeneration and tissue repair. Their biocompatibility and the fact that their surface chemistry may be modified make clay particles an ideal material for creating these mechanisms. In conclusion, the current advances in clay material research are redefining material science and stimulating cooperation between other fields of study to address global concerns. As a result of the exploration and manipulation of clay materials, transformative solutions are being produced across the spectrum of sustainability, environmental remediation, technological innovation, and healthcare. This emphasizes clay's pivotal role in advancing scientific knowledge and improving societal well-being [50].

5.2 Unveiling the multifaceted impact of clay composite material research

The research of clay composite materials is a developing topic of inquiry, defined by its tremendous potential to stimulate advances across a broad spectrum of scientific and technical fields. These materials are being methodically developed to improve their inherent qualities to meet the requirements of current applications regarding sustainability, performance, and usefulness. The advancement of clay composites underscores a concerted effort toward environmental sustainability, prioritizing materials with reduced ecological footprints [51]. Current research in this domain focuses on synthesizing composites integrating bio-based or recycled constituents, aiming to mitigate carbon emissions and energy consumption. This strategic pursuit aligns with global sustainability imperatives and underscores a dedication to ecological stewardship. Clay composites are pivotal in the construction sector for enhancing material properties such as thermal insulation, fire resistance, and moisture management. This transformative trajectory is substantiated by empirical evidence [52], showcasing the escalating adoption of clay composites in construction practices. Such adoption represents a paradigm shift in building methodologies and augments occupant safety and well-being. Moreover, as delineated in scholarly discourse [52], the integration of clay nanocomposites underscores a burgeoning trend, particularly in high-tech industries such as electronics and aerospace. These nanocomposites are meticulously engineered to exhibit exceptional mechanical and thermal attributes, amplifying their applicability in cutting-edge technological domains. Incorporating clay nanoparticles into polymers produces materials that display increased strength, flexibility, and resistance to environmental stresses. **Table 4** shows the synergy between clay composites and sustainable development, highlighting their role in construction, electronics, aerospace, and biomedicine for improved performance and environmental sustainability.

As a result, the durability and performance of electronic devices and aeronautical structures are improved. In addition, the discipline of biomedicine is investigating the possibility of employing clay composites in developing enhanced medicinal delivery systems, regenerative medicine, and tissue engineering [53]. Researchers are creating platforms for precise pharmacological drug delivery and cellular proliferation promotion, utilizing clay's biocompatibility and customizable surface characteristics. The study trajectory of clay composite materials is defined by a convergence of material science, chemistry, and engineering concepts, exemplifying a multidisciplinary research approach [54]. Clay composites are emerging as essential contributors to this endeavor, having the potential to alter industry practices, environmental sustainability, and healthcare results. This may be attributed to the intensification of the search for creative, sustainable, and high-performance materials.

Field	Key objectives	Impacts
Environmental Sustainability	Minimize ecological footprint through bio-based or recycled components, reducing carbon emissions and energy usage.	Aligns with global sustainability goals, demonstrating ecological stewardship.
Construction	Improve thermal insulation, fire resistance, and moisture management to enhance energy efficiency and safety.	Contributes to safer, more energy-efficient buildings and enhanced occupant well-being.
Electronics & Aerospace	Manufacture components with exceptional mechanical and thermal qualities using clay nanocomposites.	Increases durability and performance of electronic devices and aeronautical structures.
Biomedicine	Develop enhanced medicinal delivery systems, regenerative medicine, and tissue engineering through biocompatible clay composites.	It enables precise drug administration and promotes cellular proliferation and tissue regeneration.

Table 4.
Impacts of clay composite materials on sustainability and industry innovation.

5.3 Toward eco-efficient construction: strategies for sustainable clay-building practices

The imperative to mitigate the construction sector's environmental impact has spurred a reassessment of traditional materials. Clay has emerged as a focal point due to its minimal ecological footprint and abundant availability [55]. Strategies integrating environmental science, material engineering, and sustainable development concepts are essential for enhancing the sustainability of clay buildings. A fundamental aspect of these strategies is the thermal optimization of clay materials. The potential of clay's thermal inertia to attenuate variations in the interior environment, hence lowering reliance on mechanical heating and cooling systems, has been demonstrated through empirical research demonstrating this capacity. Advancements in composite technology, which use bio-based additives or recovered materials, can enhance heat resistance and correspond with the ideas of circular economies, which reduce waste and promote resource efficiency. The second important factor is the development of environmentally friendly methods for manufacturing bricks. Energy-intensive and carbon-emitting are two characteristics that characterize conventional kiln fire methods. On the other hand, novel initiatives, such as the use of alternative, low-energy fire systems that use solar energy or biomass, promise significant reductions in the emissions of greenhouse gases. In the context of clay construction materials, these technologies are illustrative of a change toward production paradigms that are more environmentally friendly and embrace the ideals of green manufacturing. Additionally, incorporating green building standards specifically customized to clay construction materials is essential to promote sustainable practices. By imposing stringent requirements on energy use, material sourcing, and lifetime emissions, these standards incentivize the construction sector to adopt ecologically beneficial methods. Certification programs are pivotal in cultivating a market for environmentally responsible clay-building solutions, fostering innovation and environmental stewardship within the industry [56]. The revival of traditional clay construction techniques in modern architectural practices embodies a sustainable innovation strategy. When reimagined with contemporary engineering and sustainability standards, techniques such as cob, wattle, and daub offer viable pathways to reduce the building industry's ecological footprint, characterized by their minimal environmental impact and energy efficiency. Implementing a multidimensional strategy incorporating material innovation, sustainable manufacturing processes, regulatory frameworks, and the rehabilitation of traditional methodology is required to enhance clay construction sustainability strategically [56]. This comprehensive viewpoint, supported by research from various disciplines, is of the utmost importance in increasing the ecological viability of buildings based on clay. This, in turn, contributes to the overall goals of sustainable development and environmental conservation within the built environment.

6. Conclusion

The chapter thoroughly examines the changing role of clay materials in architecture, focusing on sustainability and creativity. The process starts with a comprehensive historical analysis, emphasizing the traditional importance of clay and its subsequent evolution into a modern construction substance. The essay delves into technical advancements enhancing clay's properties, notably its heightened durability and energy efficiency. These advancements render clay an appealing choice for modern architects and builders, facilitating the creation of esthetically pleasing and environmentally sustainable structures. Additionally, the chapter explores using clay materials with other sustainable techniques, such as passive cooling and heating, demonstrating how clay enhances energy-efficient architectural designs. The text showcases many case studies that show the effective utilization of these materials in diverse architectural endeavors around the globe. These examples underscore clay's adaptability in meeting modern construction needs, like reducing carbon footprints and enhancing building performance. A comprehensive analysis examines barriers impeding clay's widespread adoption in the industry, including structural limitations, financial considerations, and regulatory hurdles. The chapter advocates for a comprehensive approach to address these problems, including regulatory reforms, enhanced research and development, and public awareness campaigns to cultivate the acceptance of clay as a feasible and desirable material in sustainable design. The chapter's conclusion serves as both an appeal to act and a contemplation on the transformative capacity of clay materials in the construction sector. The text emphasizes the significance of adopting inventive methods and fostering cooperation among architects, engineers, policymakers, and communities to use clay's capabilities fully. The essay implies that by consistently introducing new ideas and executing them in a well-planned manner, Clay may substantially contribute to attaining sustainable development objectives, especially regarding urbanization and climate change. This comprehensive synthesis highlights the necessity of a fundamental shift in construction processes, asking the industry to reevaluate conventional materials such as clay in response to contemporary technological advancements and environmental obstacles. The chapter presents a compelling case for the significant contribution that clay, with its extensive historical background and unexplored possibilities, can make in constructing a constructed environment that is both ecological, efficient, and esthetically pleasing.

Author details

Amjad Almusaed[1*], Ibrahim Yitmen[2] and Asaad Almssad[3]

1 Department of Construction Engineering and Lighting Science, Jönköping University, Jönköping, Sweden

2 Department of Construction Engineering and Lighting Science, Karlstad University, Karlstad, Sweden

3 Department of Engineering and Chemical Sciences, Karlstad University, Karlstad, Sweden

*Address all correspondence to: amjad.al-musaed@ju.se

References

[1] Khan et al. An examination of cutting-edge developments in bamboo-PLA composite research: A comprehensive review. Renewable and Sustainable Energy Reviews. 2023;**188**:113832

[2] Abyzov VA et al. Innovative building materials in the creation of an architectural environment. In: IOP Conference Series: Materials Science and Engineering. Vol. 907, No. 1. UK: IOP Publishing; 2020. p. 012035

[3] Sev A. How can the construction industry contribute to sustainable development? A conceptual framework. Sustainable Development. 2009;**17**(3):161-173

[4] Fernandes FM, Lourenço PB, Castro F. Ancient Clay bricks: Manufacture and properties. In: Dan MB, Přikryl R, Török Á, editors. Materials, Technologies and Practice in Historic Heritage Structures. Dordrecht: Springer; 2010. DOI: 10.1007/978-90-481-2684-2_3

[5] La Noce M, Lo Faro A, Sciuto G. Clay-based products sustainable development: Some applications. Sustainability. 2021;**13**:1364. DOI: 10.3390/su13031364

[6] Berman B. 3-D printing: The new industrial revolution. Business Horizons. 2012;**55**(2):155-162. DOI: 10.1016/j.bushor.2011.11.003

[7] Hussain A, Kamal MA. Energy efficient, sustainable building materials: An overview. Critical Engineering Materials. 2015;**650**:38-50

[8] Sakhdari S. Strategic integration [thesis]. USA: Virginia Tech; 2024

[9] Sharami HJ, Hosseini SJ. Theoretical framework of the Isfahani style: Inspiring sustainable aspects of a vernacular urban development. Frontiers of Architectural Research. 2024;**13**:349-369

[10] Yu L et al. Comprehensive utilization of blast furnace slag, municipal sludge, and kaolin clay in building brick manufacture: Crystalline transformation, morphology observation, and property assessment. Cement and Concrete Composites. 2024;**145**:105337

[11] Garcez LR et al. Characteristics of the açai seed (Euterpe precatoria Martius) after thermal processing and its potential in soil-cement brick. Case Studies in Construction Materials. 2024;**20**:e02816

[12] Schleck G, Ben-Alon L. Eco-ableism and access circularity in natural building. Frontiers of Architectural Research. 2024;**13**:235-248

[13] Kanagaraj B et al. Limestone calcined clay cement (LC3): A sustainable solution for mitigating environmental impact in the construction sector. Advances Resources, Conservation & Recycling. 2024;**21**:200197

[14] Kandpal V et al. Sustainable energy transition, circular economy, and ESG practices. In: Sustainable Energy Transition: Circular Economy and Sustainable Financing for Environmental, Social and Governance (ESG) Practices. Cham: Springer Nature Switzerland; 2024. pp. 1-51

[15] Pazienza A, Baselli G, Vinci DC, et al. A holistic approach to environmentally sustainable computing. Innovations in Systems and Software Engineering. 2024. DOI: 10.1007/s11334-023-00548-9

[16] Shehawy YM, Faisal SM, Khan A. Consumer readiness for green

consumption: The role of green awareness as a moderator of the relationship between green attitudes and purchase intentions. Journal of Retailing and Consumer Services. 2024;**78**:103739

[17] Diouf G. Millenium development goals (MDGs) and sustainable development goals (SDGs) in social welfare. International Journal of Science and Society. 2019;**1**(4):17-24

[18] Valencia SC et al. Adapting the sustainable development goals and the new urban agenda to the city level: Initial reflections from a comparative research project. International Journal of Urban Sustainable Development. 2019;**11**(1):4-23

[19] Samer M. Towards implementing the green building concept in agricultural buildings: A literature review. Agricultural Engineering International: CIGR Journal. 2013;**15**(2):25-46

[20] Lodi L. Developing a sustainable circular economy system in the roofing and geo composites industries [thesis]. Italy: Politecnico di Torino; 2023

[21] Gabriel M, Luque MLD. Sustainable development goal 12 and its relationship with the textile industry. In: The UN Sustainable Development Goals for the Textile and Fashion Industry. UK: Springer; 2020. pp. 21-46

[22] Bergman D. Sustainable Design: A Critical Guide. New York: Princeton Architectural Press; 2013

[23] Al-mused A et al. Shading effects upon cooling house strategy in Iraq. In: 2nd PALENC Conference and 28th AIVC Conference on Building Low Energy Cooling and Advanced Ventilation Technologies in the 21st Century. Greece: University of Athens; 2007

[24] Almusaed A, Almssad A, Yitmen I. Sustainable Built Environment and its Implications on Real Estate Development: A Comprehensive Analysis. London, UK: IntechOpen; 2024

[25] Almusaed A, Almusaed A. The urban heat island phenomenon upon urban components. In: Biophilic and Bioclimatic Architecture: Analytical Therapy for the Next Generation of Passive Sustainable Architecture. UK: Springer Science & Business Media; 2011. pp. 139-150

[26] Al-Dmour. Youmna—flourishing the indoor environment quality of workplaces using the biophilic architectural design [thesis]. UK: Brunel University London; 2023

[27] Zhong W, Schröder T, Bekkering J. Biophilic design in architecture and its contributions to health, well-being, and sustainability: A critical review. Frontiers of Architectural Research. 2022;**11**(1):114-141

[28] Gillis K, Gatersleben B. A review of psychological literature on the health and wellbeing benefits of biophilic design. Buildings. 2015;**5**:948-963. DOI: 10.3390/buildings5030948

[29] Labonnote N et al. Additive construction: State-of-the-art, challenges, and opportunities. Automation in Construction. 2016;**72**:347-366

[30] Almusaed A, Almssad A. Introductory chapter: Bricks between the historical usage and sustainable building concept. In: Masonry for Sustainable Construction. London, UK: IntechOpen; 2023

[31] Jaglarz A. Perception of color in architecture and urban space. Buildings. 2000;**2023**:13. DOI: 10.3390/buildings13082000

[32] Almusaed A, Almssad A. Urban biophilic theories upon reconstruction processes for Basrah City in Iraq. In: Passive and Low-Energy Architecture Conference. Switzerland: PLEA; 2014

[33] Almssad A, Almusaed A. Efficient daylighting approach using light-shelve device adequate for habitat program in Aarhus City. International Journal of Smart Grid and Clean Energy. 2014;**3**(4):441-453

[34] Al-Kodmany K. Greenery-covered tall buildings: A review. Buildings. 2023;**13**:2362. DOI: 10.3390/buildings13092362

[35] Hallsworth JE, Udaondo Z, Pedrós-Alió C, Höfer J, Benison KC, Lloyd KG, et al. Scientific novelty beyond the experiment. Microbial Biotechnology. 2023;**16**(6):1131-1173

[36] Carr D. Moral values and the arts in environmental education: Towards an ethics of aesthetic appreciation. Journal of Philosophy of Education. 2004;**38**(2):221-239

[37] Kanaani M, editor. The Routledge Companion to Paradigms of Performativity in Design and Architecture: Using Time to Craft an Enduring, Resilient, and Relevant Architecture. 1st ed. London: Routledge; 2024

[38] Yassine MB. Contemporary architecture design of clay material-ecological alternatives for individual houses in the Sahara desert. Materials Science Forum. 2020;**1005**:57-64. DOI: 10.4028/www.scientific.net/msf.1005.57

[39] Jones RII. The master potter and the rejected pots: Eugenic legislation in Victoria, 1918-1939. Australian Historical Studies. 1999;**29**(113):319-342

[40] Almssad A, Almusaed A, Homod RZ. Masonry in the context of sustainable buildings: A review of the brick role in architecture. Sustainability. 2022;**14**(22):14734

[41] Habert G et al. Environmental impacts and decarbonization strategies in the cement and concrete industries. Nature Reviews Earth & Environment. 2020;**1**(11):559-573

[42] Almusaed A, Almssad A, Alasadi A, Yitmen I, Al-Samaraee S. Assessing the role and efficiency of thermal insulation by the "BIO-GREEN PANEL" in enhancing sustainability in a built Environment. Sustainability. 2023;**15**:10418. DOI: 10.3390/su151310418

[43] Chen Y et al. Intelligent designs from nature: Biomimetic applications in wood technology. Progress in Materials Science. 2023;**139**:101164

[44] Davies GM, Kettridge N, Stoof CR, Gray A, Ascoli D, Fernandes PM, et al. The role of fire in UK peatland and moorland management: The need for informed, unbiased debate. Philosophical Transactions of the Royal Society B: Biological Sciences. 2016;**371**(1696):20150342

[45] Leifeste A, Stiefel BL. Sustainable Heritage: Merging Environmental Conservation and Historic Preservation. UK: Routledge; 2018

[46] Andrew JJ, Dhakal HN. Sustainable biobased composites for advanced applications: Recent trends and future opportunities–a critical review. Composites Part C: Open Access. 2022;**7**:100220

[47] Landi DA, Frota F, et al. Life cycle assessment of a novel fired smart clay brick monitoring system for masonry buildings. Sustainable

Energy Technologies and Assessments. 2022;**50**:101745

[48] Muntari MY, Windapo AO. Clay as sustainable building material and its benefits for protection in the built Environment. In: IOP Conference Series: Materials Science and Engineering. Vol. 1144, No. 1. UK: IOP Publishing; 2021

[49] Gul S et al. Research progress on properties and applications of polymer/clay nanocomposite. Polymer-Plastics Technology and Engineering. 2016;**55**(7):684-703

[50] Jørgensen MS et al. Green Technology Foresight about Environmentally Friendly Products and Materials. Denmark: Danish Environmental Protection Agency; 2006. Available from: http://www2.mst.dk/udgiv/publications/2006/87-7052-216-2/pdf/87-7052-217-0.pdf

[51] Kauschen JS. Life Cycle Thinking and Integrated Product Deliveries in Renovation Projects: Extending the Concept of Integrated Product Deliveries with Product Service Systems. Germany: The Centre for Sustainable Design (CfSD)/UCA–University of Creative Arts, Farnham, UK; 2012

[52] Casini M. Smart Buildings: Advanced Materials and Nanotechnology to Improve Energy-Efficiency and Environmental Performance. UK: Woodhead Publishing; 2016

[53] Gaharwar AK et al. 2D nanoclay for biomedical applications: Regenerative medicine, therapeutic delivery, and additive manufacturing. Advanced Materials. 2019;**31**(23):1900332

[54] Otaiku AA. Significance of value-added material. International Journal of Current Research. 2019;**11**(10):7891-7907. DOI: 10.24941/ijcr.33599.10.2019

[55] Environment UN et al. Eco-efficient cement: Potential economically viable solutions for a low-CO_2 cement-based materials industry. Cement and Concrete Research. 2018;**114**:2-26

[56] Appolloni A et al. Green recovery in the mature manufacturing industry: The role of the green-circular premium and sustainability certification in innovative efforts. Ecological Economics. 2022;**193**:107311

Chapter 2

Nanoclays as Fillers for Performance Enhancement in Building and Construction Industries: State of the Art and Future Trends

Zenamarkos Bantie, Asmare Tezera, Desalegn Abera and Tesfa Nega

Abstract

In construction engineering, there is currently a strong emphasis on finding construction materials, mainly the binder which plays a crucial role, that meet multiple criteria, including sustainability, cost-effectiveness, durability, and reduced environmental impact. However, there is a growing interest in exploring alternatives to traditional binders to address the limitations associated with their production and use. One such alternative is the use of naturally occurring materials like clay. Clay deposits are abundant and widely available, making them a sustainable resource for construction applications. Moreover, clay contains significant amounts of silica and alumina, which are key components for inducing pozzolanic reactions that contribute to the strength and durability of concrete. In recent studies, nanoclays (NCs) have emerged as a promising addition to construction materials as supplementary cementitious materials. These nanoparticles possess unique properties that can enhance the performance of concrete. Nanoclays significantly improve the compressive strength, sustainability, and durability of concrete structures. The high surface area and reactivity of nanoclays facilitate better bonding between cement particles, resulting in enhanced mechanical properties. This chapter aims to discuss the state of the art on performance enhancements of building materials that employ different types of nanoclays in place of conventional binders and the future trends.

Keywords: nanoclays, cement, bitumen, geopolymers, sustainable construction

1. Introduction

Clay is a natural, fine-grained mineral that hardens upon drying and plastic when moist, containing phyllosilicates and other substances that provide plasticity, along with additional phases and organic matter [1, 2]. Clay properties, including mineralogy and composition, are influenced by formation conditions, parent rock characteristics, and chemical transformations in volcanic ash [3, 4]. Clay minerals, originating from earth's crust and atmosphere, have unique characteristics like structure, fine particle size, and cation-exchange capacity, affecting water absorption, swelling, impermeability, and plasticity [5]. Clay, derived from the Greek word "argilos," is a fine-grained material with plasticity and hardening properties, formed through weathering, the breakdown, and chemical decomposition of igneous rocks [6]. Clay particles are formed through the long-term weathering and erosion of rocks containing soil, ceramic clays, clay shales, glacial clays, and significant amounts of detrital and transported clays [7, 8]. Clays and clay minerals form in various geological environments like soil horizons, sediments, geothermal fields, volcanic deposits, and weathering rock formations through physical breakdown and chemical transformation [9, 10]. Clays are extensively utilized in various industries, including refractory manufacturing, pottery/ceramic production, structural unit fabrication, and fillers or extenders in various products [11]. Ancient civilizations valued clay materials in construction, pottery, brick making, ceramics, paints, plastics, rubbers, cosmetics, and pharmaceuticals, as they enhanced the performance and properties of materials. **Figure 1** [3, 12, 13]: as depicted below shows formation of clays.

As shown in **Figure 2**, clay minerals are categorized by their chemical environment at the interface between rocks and the atmosphere, leading to destabilization and subsequent burial under high temperatures [14].

Clays are fine-grained 2D earthen minerals with a chemical composition of 60% silica and 15% alumina, with other mineral oxides in the remaining fraction [15]. Clay minerals in soil indicate site weathering and deposition history, providing insights into geological processes in residual soils originating from local rock [16].

In the field of soil mechanics, some of the key properties of clays include hydraulic conductivity, gas conductivity, ion diffusion capacity, swelling potential, compressibility, and rheological properties [17]. Clay minerals in soil indicate site weathering

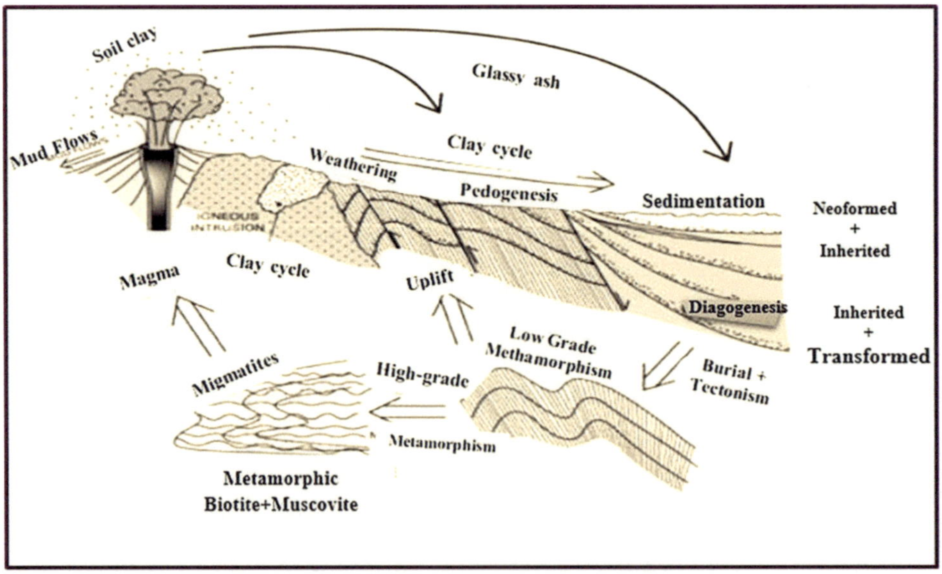

Figure 1.
Process of clay formation [3].

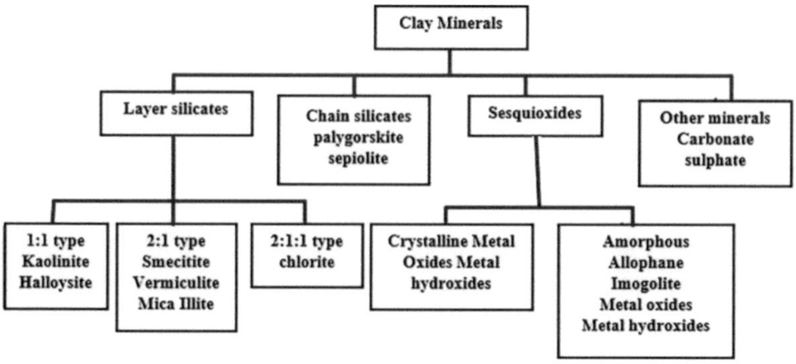

Figure 2.
Classification of clay minerals [7].

Primary properties	*Secondary properties*
Chemical composition	Granulometric composition
Density	Adsorptive capacity
Hardness	Specific density
Surface properties	Abrasiveness
Color	Precipitation behavior
Wettability	Rheological performance
Plasticity	Swelling capacity
Cation exchange capacity	Pore structure

Table 1.
Primary and secondary properties of clay [18].

and deposition history, providing insights into geological processes in residual soils originating from local rock [18]. **Table 1** as shown below provides a comprehensive overview of the primary and secondary properties associated with clay.

Clay nanoparticles are attracting interest due to their abundant natural availability, accessibility, effectiveness, and environmentally friendly characteristics. With a small grain size and unique crystal structure, they possess exceptional physicochemical properties such as a large specific surface area, surface charge, and cation exchange capacity [19, 20]. Clay minerals are layered silicates, composed of stacked aluminosilicate layers with high surface area, formed by fused tetrahedral and octahedral sheets and unshared oxygen atoms in hydroxyl form [21].

Research shows that the physical properties of modified cementitious materials are significantly influenced by factors like strength, porosity, mix composition, water content, specimen geometry, nanoparticle type, and age [22]. Consistency in cohesive soils is crucial for their strength, resistance to penetration, stability, and load-bearing capacity, influencing construction and geotechnical engineering practices [23]. The phyllosilicate structure's fundamental unit is an aluminosilicate layer consisting of a silica tetrahedral sheet and an alumina octahedral sheet, combined in specific proportions [24]. Nanoclay fillers like carbon nanotubes, graphene, nanocellulose, and nanosilica are renowned for their reinforcing properties, with clay being a notable filler due to its composition of layered silicates/clay minerals [25]. Nanotechnology enhances concrete properties by preventing cracks, reducing shrinkage, minimizing joint need, mitigating temperature curling, and reducing moisture loss warping [26]. Nano-sized silicon dioxide/nanosilica can enhance concrete properties by inhibiting water penetration, increasing density, and reducing environmental impact associated with concrete production and use [27]. Nanoparticles, including titanium dioxide, carbon nanotubes, silica, copper, clay, and aluminum oxide, have the potential to significantly improve the physical and chemical properties of construction materials [28]. Nanomaterials in bridges, composites, reservoirs, roads, and buildings offer benefits like reduced maintenance, thermal transfer, sound absorption, improved glass reflectivity, corrosion resistance, and lower life-cycle costs [29]. Nanosilica improves the interface between binding paste and concrete aggregates, enhancing mechanical strength, reducing water absorption, and sorptivity due to fewer capillary pores and pozzolanic activity [30].

Nanoparticles have been used in cement-based materials to enhance the performance and durability of the concrete [31]. Nanoclays, also known as layered silicates, are widely employed as nanofillers in the production of polymer layered silicate nanocomposites. Phyllosilicates (2:1), a type of layered silicate, are particularly favored and extensively utilized in the preparation of clay-based nanocomposites [32]. The incorporation of nanoclay into concrete has gained worldwide recognition in recent years, owing to its immense potential advantages and its growing significance within the construction sector. The attention given to clay minerals is justified by their abundance, accessibility, effectiveness, and ecological properties and have a unique crystal structure, are recyclable, easy to use, and possess remarkable physicochemical properties such as specific surface, surface charge, and cation-exchange capacity [33].

2. Synthesis and modification of nanoclays

2.1 Clay minerals

Clay, a hard earth material, can significantly alter the chemical and isotopic compositions of solid and fluid phases during weathering due to its unique layered structure and potency as an ion exchanger [34]. It consists of mainly fine particles of hydrous aluminum silicates and other minerals [35]. They can be found in many different types of habitats, including deep-sea sediments, lakes, estuaries, soils, weathering profiles, and hydrothermal systems [36]. Clays are typically made up of a diverse array of minerals, often mixed with other minerals and organic materials like feldspar, quartz, calcite, and organic matter [36–38]. Although these contaminants are important to the physicochemical characteristics of clays, their presence in high concentrations has a detrimental effect on clays' performance [3]. They are used in a number of industries such as biomedical [39], health care [40], construction [28, 41–44], environmental remediation, food packaging [45, 46], clothing and paints [47], soil fertility [48], etc. This chapter only focuses on the advancements of clay minerals (nanoclays) in construction industries.

2.2 Synthesis and modification of nanoclays

Nanoclay is a component of phyllosilicates, mainly defined by the presence of silicon, oxygen, and other elements. It comes from natural sources and has undergone subsequent chemical processing [48]. Nanoclay is made up of thin layers, each of which is between one and several nanometers thick and between several hundred and several thousand nanometers long [49]. There are a number of methods for synthesizing nanoclay, such as *in-situ* polymerization, melt-blending, and solution-blending [50]. The solution-blending approach is combining a polymer solution with nanoclay and then letting the solvent evaporate [51]. The melt-blending technique entails combining liquid polymer and nanoclay, then cooling and solidifying the mixture [52]. The *in-situ* polymerization method is the polymerization of monomers in the existence of nanoclay, leading to the formation of nanoclay-polymer composites [53]. Selecting the right synthesis approach for nanoclays depends on desired properties, such as scalability, cost-effectiveness, and compatibility with the polymer matrix.

Naturally occurring clay minerals include mixed cations on their surface and in the interlayer space, which makes their surface characteristics unsuitable for use in some applications [38]. To give clay minerals their unique physicochemical properties for definite purposes, their interlayer space is saturated with the required cations. The modification of clay particles can be done through (i) physical and (ii) chemical modification methods [54]. Physical modification of clay surface by modifying substances marginally improves polymer composite characteristics without altering clay structure due to minimal physical forces. Chemical modification of clay surfaces with tethered polymers or organo-silane compounds improves properties of polymer/clay nanocomposite systems, including improved dispersion, nano-sized layers, high aspect ratio, and large surface area [55]. Surface modification can be achieved by various methods, including acid treatment, surfactant treatment, and polymer grafting. Acid treatment involves the use of acid to remove impurities and increase the surface area of nanoclay [50]. Surfactant treatment involves the use of surfactants to modify the surface chemistry and improve the dispersion of nanoclay in polymers [56]. Polymer grafting involves the attachment of polymer chains to the surface of nanoclay, leading to improved compatibility with polymers [55, 57].

3. Structure and properties

Understanding the clay structure and its characteristics is crucial for comprehending the impact of clay reinforcement on the performance enhancement of construction and building materials. Nanoclays are layered mineral silicates composed of individual layers of octahedral ($[AlO_3(OH)_3]6$ or tetrahedral sheets $(SiO_4)4$ [58–60], stacking on top of one another like pages in a book [61]. In tetrahedral sheets, a silicon atom is surrounded by four oxygen atoms, and in octahedral sheets, a metal like magnesium or aluminum is surrounded by eight oxygen atoms, and are the building blocks of clay's layers. Tetrahedral (T) and octahedral (O) sheets are linked by oxygen atoms [3, 62, 63], while unshared oxygen atoms are present in hydroxyl form [62, 64, 65]. The octahedral sheet is formed by four apical oxygen atoms and two hydroxyl groups, with cavities occupied by trivalent or bivalent cations. Aluminum is typically used, and the structure is called di-octahedral when two cavities are trivalent cations. Tri-octahedral is formed when all cavities are divalent metal ions. Isomorphic substitutions can balance the charge deficit [3, 38, 62, 66–69]. In clay crystallite, multiple layers are joined by hydrogen bonds, Vander Waals force, and electrostatic force to form sheets, forming stacks of parallel lamellae [32, 61, 70]. Clay minerals consist of four structural types: three layered (1: 1, 2: 1, 2: 1: 1) and one fibrous structure (**Figure 3**) [3, 38, 62, 68, 71]. In 1:1 arrangements, the layer is

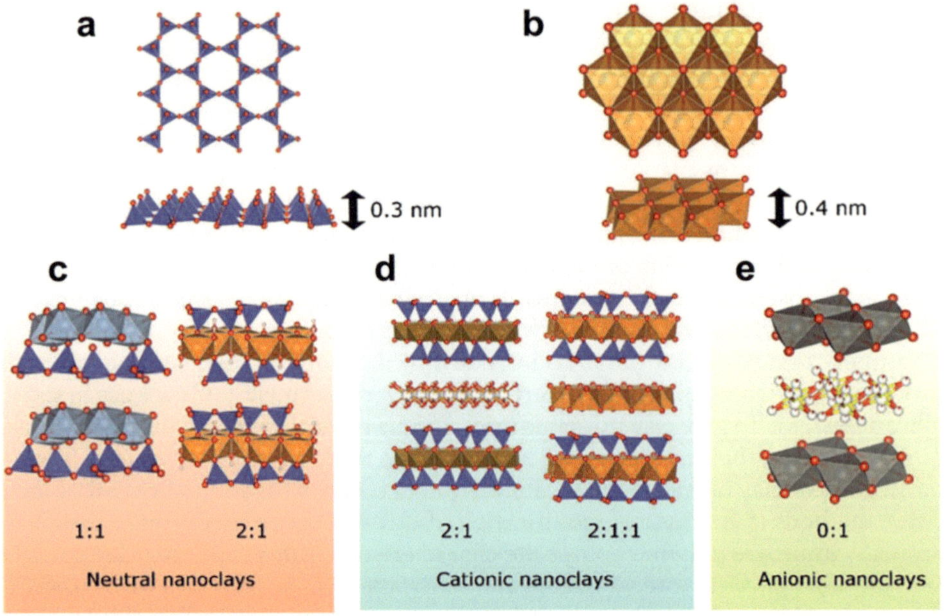

Figure 3.
Schematic representation of (a) tetrahedral, (b) octahedral sheet; (c) layered nanoclay 1: 1, and 2: 1 structures; (d) cationic nanoclays (types 2: 1 and 2: 1: 1), and (e) anionic nanoclay (type 0: 1) sheet arrangements in common nanoclay materials [68].

formed by one tetrahedral sheet linked to one octahedral sheet; in 2:1 arrangements, each octahedral sheet is connected to two tetrahedral sheets (one sheet on each side); and in 2:1:1 configurations, each octahedral sheet is adjacent to another octahedral sheet and connected to two tetrahedral sheets [64, 72]. Clay minerals with fibrous structure consist of ribbon-like layers of two tetrahedral sheets held together by a central octahedral sheet through shared oxygen; which results in a gutter-and-channel-type structure [73]. Depending on the arrangement of layers, bond types, chemical composition, and microscopic structure of the crystal lattice, there are different types of nanoclays, including montmorillonite, bentonite, kaolinite, hectorite, and halloysite nanoclay [74–76].

Kaolinite is a 1:1 type of clay mineral with a silica tetrahedral sheet connected to an alumina octahedral sheet connected by hydrogen atoms, with the chemical formula Al2Si2O5 (OH)4 [77–79]. The nanoclay of the kaolinite group has remarkable chemical and physical characteristics, including a high specific surface area, corrosion resistance, low density, and high thermal stability [79]. In natural kaolinite, Al^{3+} is replaced by Ca^{2+}, Mg^{2+}, or Fe^{2+}, while Si^{4+} is replaced by Al^{3+} or Fe^{3+}, resulting in an insufficient charge in the crystal lattice and a negatively charged surface [60]. Montmorillonite (MMT) nanoclay has a 2:1 structure with weak Van der Waals forces keeping the outer layers together, consisting of $(Na,Ca)_{0.33}(Al,Mg)_2(Si_4O_{10})(OH)_2{\cdot}nH_2O$ chemical composition [77, 80]. MMT, with its high aspect ratio, surface area, high cation-exchange capacities, and excellent mechanical properties, can swell up to several times its original size when exposed to water [48, 81]. Al^{3+} and Mg^{2+} are isomorphic replacements of Si^{4+} in the tetrahedral sheet and Al^{3+} in the octahedral sheet, respectively, to form the chemical compound known as MMT [59, 82]. Halloysite, similar to kaolinite in chemical composition, differs from kaolinite due to

Clay	Specific surface areas (m^2/g)	Cation exchange capacity (CEC) (meq/100 g)
Kaolinite	5-40	3-15
Halloysite (hydrated)	1100*	40-50
Illite	10-100	10-40
Smectites	40-800	80-120
Vermiculite	760*	100-150
Chlorite	10-55	10-40
Palygorskite-sepiolite	40-180	3-20
Allophane	2200	30-135
Imogolite	1540	20-30

**depending on the fraction of internal specific surface area.*

Table 2.
Cation-exchange capacities and specific surface areas of clay minerals [38, 86].

its unique structure consisting of a single water molecule layer with chemical composition $Al_2Si_2O_5(OH)_4.nH_2O$) [79]. Halloysite is characterized by 15-20 thick layers, forming a hollow and porous tubular structure with hydrophilic and negative charges, strong adsorption, high specific surface area, water dispersion, biocompatibility, and active surface chemical [79, 83]. Some of the better-known members of the clay family have the following chemical formula: talc ($Mg_3[Si_4O_{10}(OH)_2]$), mica(KA_{l2}[Al $Si_3O_{10}(OH)_2$]), serpentine, ($Mg_3[Si_2O_5(OH)_4]$), sepiolite(Mg4$[Si_6O_{15}]$ $(OH)_2.4H_2O$) [68, 84]. The chemical composition of nanoclays may be different, including the aluminum silicates, magnesium silicates, aluminum–magnesium silicates, and other mineral silicate nanoforms [85]. **Table 2** depicted the cation-exchange capacities and specific surface areas of clay minerals [38, 86].

4. Nanoclays for the performance enhancement in building and construction materials

4.1 Mechanical properties

4.1.1 Nanoclays in concrete and asphalt binders

Various types of nanomaterials (such as nano-silica, carbon nanotubes, nano-titanium dioxide, nano-alumina, nanoclay, nano-zinc oxide,) are employed to enhance the strength, durability, and overall physical and chemical properties of concrete [87]. The use of nanoparticles in construction enables more cost-effective, time-efficient, and safer production of construction materials, resulting in enhanced product durability, efficiency, and higher performance levels in the production of raw materials [88]. The incorporation of different nanoparticles into concrete at appropriate concentrations has the potential to enhance its mechanical strength and chemical durability, and reduce shrinkage, while also minimizing issues such as bleeding and segregation [89]. The introduction of nano-sized particles fills the interfacial transition zone between cement and aggregate, leading to densification of the cement matrix by enhancing the formation of calcium-silicate-hydrate (C-S-H) gel through the pozzolanic reaction with calcium hydroxide [90]. Nano particles also have the capability to enhance the bond between aggregates and cement paste [91]. Nanoparticles are highly effective additives for modifying cement products, even when used at low concentrations of approximately 1% [92]. Another research [93] demonstrated that nanoclay admixture improved the compressive property of cement calcareous sand (CCS) with an optimum admixture ratio of 8%, enhanced the deformation modulus and resistance to external load deformation, and increased the density of the internal structure, thereby improving the mechanical properties of CCS.

In the pavement industry, natural nanoclays, particularly montmorillonite, hectorite, saponite, and kaolinite, are commonly used, while organically modified montmorillonite known as Cloisite is utilized as an asphalt modifier [94]. Cloisite-modified asphalt mixture, nano clay modification was found to enhance the mechanical properties, including creep resistance, tensile strength, and fatigue resistance, effectively improving the performance of the polymer as a filler reinforcement [95]. The utilization of nanomaterials in the construction industry holds immense potential for enhancing the performance and properties of building materials [96]. The high aspect ratio and large specific surface area of nanoclays contribute to improved load transfer, enhanced interfacial bonding, increased resistance against cracking, improved flexural and compressive strength, and enhanced impact resistance [97]. Nanoclays, offer a promising alternative to traditional additives, providing exceptional mechanical properties and unique geometrical characteristics. Their precise distribution fills nano-sized voids in cement mortar, limiting micro-crack propagation and improving strength, workability, consistency, and water absorption [98]. The incorporation of nanoclay into concrete has several beneficial effects, including reducing building energy consumption, enhancing thermal insulation properties, improving thermal stability, fire resistance, thermal performance, behavior, resistance to cracking and degradation, and decreasing thermal conductivity, which leads to effective insulation and lower energy consumption during heating and cooling periods [33]. Nanoparticles contribute to the development of strength in cement-based materials by enhancing the hydration process of cement, resulting in the formation of calcium-silica-hydrate (C-S-H) and improved overall strength [99].

4.1.2 Nanclays in polymer composites

Nano-particles are currently recognized as highly promising fillers for enhancing the mechanical and physical properties of composite materials [100]. The incorporation of hydrophilic clay minerals into hydrophobic polymers can enhance their mechanical performance, fire retardant, and barrier resistance, as well as their thermal and electrical conductivities [101]. Adding nanoclay particles below or above their optimum point reduces material performance by introducing voids that act as preferential locations for crack initiation and failure [102]. Nanoclay particles are currently recognized as promising fillers for improving the mechanical and physical properties of polymers [103]. The addition of fillers to polymer matrices provides an alternative approach to enhance the mechanical performance, processability, and cost-effectiveness of composites as well as reduces available free spaces, resulting in increased stiffness of the laminates and improved interaction between the fibers and the matrix [104]. According to the study [105], the experimental evidence demonstrates that increasing the weight percentage of nanoclay in Coccinia Indica fiber reinforced epoxy composites enhances their tensile, flexural, impact, and compression properties. The results found in [106] reveal that Polymethylmethacrylate nanocomposites with a minimal loading of 0.5 wt% well-dispersed montmorillonite nanoclay, significant enhancements were observed in the mechanical properties of the nanocomposites, including improvements in elastic modulus, tensile strength, and elongation at break. Based on the study conducted by [107], the incorporation of nanoclay particles in composite films resulted in a significant improvement in flexural modulus, especially at lower weight percentages.

Recent research, as referenced in [32], have documented that the fabrication and characterization of nanoclay-based nanocomposites and bio-nanocomposites, exhibited superior mechanical and thermal properties when compared to nanoclay-reinforced polymer composites. The incorporation of layered silicates into polymer nanocomposites leads to notable improvements in their mechanical properties, including modulus, strength, toughness, and strain [108]. Nanoparticles enhance the properties of many polymers and are used to improve modulus, tensile strength, flame resistance (to further improve the flame retardant performance), and thermal and structural properties in various applications [109]. Nanoparticles can serve as heterogeneous nuclei in cement pastes, effectively accelerating cement hydration due to their high reactivity. They also act as nano-reinforcements and nano-fillers, leading to densification of the microstructure and reducing porosity in cement-based materials [110]. As mentioned in the study referenced [111], incorporating nanoclay into the cementitious mixture for 3D printing concrete resulted in a decrease in workability but an increase in the yield stress due to enhanced cohesiveness. Increasing the amount of nanoclay not only maintained the desired shape during and after extrusion but also improved layer stability, preventing collapse under the weight of subsequent layers. Another finding reported in Ref. [112], is the successful incorporation of clay into various industries, including transportation, construction, packaging, and medical sectors, offers improved mechanical properties and enhanced functional characteristics such as fire retardancy, ablation performance, and gas permeability. In another study [113], it was observed that the addition of nanoclay to epoxy resin resulted in a significant increase in the elastic modulus and fracture toughness. However, as the nanoclay level increased, there was a notable reduction in the failure strength and failure strain. In line with the conclusions drawn in Ref. [114], the addition of a small amount of organoclay, around 4 wt%, to polymers leads to significant improvements in barrier properties, dimensional stability, flame retardant, mechanical properties such as tensile strength, Young's modulus, and flexural modulus this allows for the creation of lighter and more transparent composites.

The study outlined in Ref. [115], highlights that nanoparticles-reinforced polymer composites, particularly montmorillonite/epoxy nanocomposites, have undergone thorough investigation and have shown substantial improvements in mechanical, thermal, and physical properties. The incorporation of nanoclay into composites significantly enhances the mechanical properties (tensile strength, flexural strength, compression strength, and impact resistance) thermal stability, and flame-retardant characteristics [116]. Nanomaterials aim to enhance mechanical and physical properties at the nanoscale, improving durability [13]. The incorporation of clay in nanocomposites leads to significant enhancements in interfacial interactions with the matrix and physical properties. Based on the findings obtained in Ref. [101], the results indicate that the mechanical properties of the composites improved with an increase in filler content, and the optimum mechanical properties in percentage were found to be 4%.

Incorporating an appropriate amount of clay content into materials has been found to enhance several key properties, including tensile strength, modulus (stiffness), thermal stability, elasticity, and heat deflection temperature [117]. The incorporation of filler nanoparticles in the resin matrix of dental composites enhances their esthetic, optical, and mechanical characteristics, including factors like tensile strength, fracture resistance, and reduced polymerization shrinkage [118]. Based on the findings reported by [119], it was concluded that the addition of fillers increased the stiffness of the polymer composite. However, when the filler ratio was increased beyond a certain point, particle agglomeration occurred, leading to decreased adhesion between the matrix and fillers, resulting in a decline in the mechanical strength of the composite [119].

4.2 Thermal properties

4.2.1 Nanoclay in modification of concrete

The thermal response of concrete is crucial in construction industries. This property can be altered by using different additives which can give thermal stability to the concrete. Researchers have investigated the effect of nanoclays on the thermal stability of concrete [120–124]. Hakamy et al. [125] carried out the thermal stability of nanoclay (Cloisite 30B) and calcined nanoclay-cement nanocomposites. They found that thermal stability was improved by 3.3% compared to the control cement paste and also showed that adding Cloisite 30B nanoclay reduced the weight loss of cement mortar. The thermal stability of hemp fabric-reinforced nanoclay-cement was also investigated in another work [122]. The study found that adding nanoclay to hemp fibers-reinforced nanocomposites increased their thermal stability, as revealed by thermogravimetric analysis. The thermal behavior of cement mortar treated with kaolinite nanoclay was studied by Fan et al. [126]. According to their DTA data, the specimens' weight loss was decreased by 5% when nanoclay was added, which resulted in exceptional heat stability and resistance to heat. Mohammad and Mohammed [123] investigated the thermal performance and fire resistance of nanoclay-modified cementitious materials using montmorillonite nanoclay as partial replacement for cement. The addition of nanoclay significantly reduced the degradation in the tensile and flexural strengths of cement mortar due to elevated temperature exposure. Research work [121] also demonstrated that partial replacement of cement up to 3%wt. could reduce the thermal conductivity of the concrete considerably.

4.2.2 Nanoclay in modification of polymers

Polymers have many more uses besides concrete, prefabricated structural elements, and strengthening elements. Some of these include pipeline sheaths, plastic films, decorative panels, floors, doors, windows, and sanitary facilities. [127]. The addition of nanoclay fillers in polymer matrix have great effect in improving the thermal properties. High density polyethylene (HDPE), one of the polymer used in construction industries, can be modified with nanocomposites produced industrially by melt mixing approach [128]. Harun et al. [128] modified HDPE with nanoclay reinforcement, enhancing mechanical and thermal properties, resulting in increased glass transition temperature and decreased Vicat softening temperature. Based on exfoliated organoclay polystyrene and poly(methyl methacrylate) (PMMA), a study [129] assessed the thermal characteristics of nano montmorillonite. TGA analysis and calcination reveal that nanoclay addition enhances thermal stability in organoclay polymers, with a rise in degradation temperature of 25-50°C above neat PS and 25-80°C above neat PMMA.

4.2.3 Nanoclay in modification of asphaltic bitumen

Improvements in the pavement sector, including heavy axle design, traffic leveling, huge truckloads, and environmental requirements, necessitated the need for asphalt binders to be improved [130]. Polymer modification has become increasingly popular in recent decades as a means of enhancing asphalt pavement's high-temperature performance [131] without compromising its low-temperature behavior. One of the technique to improve high-temperature performance is using nanoclays as a reinforcement [130, 132, 133]. Atefeh et al. [134] tried to modify bitumen by ethylene propylene diene monomer (EPDM) blended with hybrid nanoparticles (carbon nanotubes (CNTs) master batch and bentonite nanoclay) to improve thermal and other properties. Thermogravimetric measurements showed that the hybrid samples significantly improved bitumen's viscoelastic behavior at high temperatures by adding minimal amounts of additives. Bahadır et al. [135] assess the thermal properties of nanoclay-modified bitumen using organo- montmorillonite nanoclay. In the DSC result the glass transition temperature of modified bitumen has increased with the application of nanoclays. Moreover, the use of nanoclay in bitumen has turned exothermic process of bitumen melting and crystallization into endothermic. The viscosity of the material at elevated temperatures surpasses that of the original binder, indicating greater resistance against permanent deformations.

4.2.4 Nanoclay in modification of geopolymer

Assaedi et al. [136] utilized nanoclay platelets (Cloisite 30B) for the mechanical and thermal properties improvements of fly ash geopolymer. The thermal stability of samples was determined using TGA. In this test, the thermal stability was studied in terms of the weight loss percentage as a function of temperature in Argon atmosphere. The neat geopolymer exhibits significant porosity, reducing water capacity, but the sample with 2% NC shows the highest thermal stability, with a nanocomposite peak slightly higher than other samples. Khater et al. [137] investigated the effect of nanoclay on alkali-activated water-cooled slag geopolymer. Kaolin nanoclay, prepared through firing at 800°C, was added to geopolymer with a 0 to 7% dry weight ratio, increasing the ratio of nanoclay from 1 to 1.5% improving thermal and thermal insulation qualities.

4.3 Rheological properties

4.3.1 Nanoclay in mofication of concretes

Rheology describes how a material deforms and flows when applied stresses and shear rates are applied. The flow curves that relate yield stress, shear rate, and plastic viscosity provide a better explanation for the flowability of fresh concrete [138]. Over the past few decades, it has been thoroughly investigated in cement-based materials, leading to the establishment of several analysis approaches. Nanoclays are widely considered to require minimal pre-dispersion in concrete production due to their ease of dispersion under shearing by conventional mixing [139]. Small amounts of nanoclay (less than 1% by binder weight) increase the green strength of self-associating concrete immediately after casting without significantly altering flowability [140, 141]. Dejaeghere et al. [140] studied the impact of nanoclays on the rheological properties of cement-based mortars. The outcome demonstrates that the card-house-like microstructure created by the extremely small, strong nanoclay particles raised the static yield stress. A study [141] on shear rheology revealed that nanoclays have minimal long-term effects and an instant stiffening effect controlled by flocculation, rather than water adsorption. The research [142] on Self-Compacting Concrete (SCC) utilized nanoclays attapulgite, bentonite, and sepiolite NC, and found that sepiolites significantly impacted the rheological characteristics of fluid cement pastes.

Figure 4a illustrates the stress development of fresh cement pastes under 0.01 s^{-1}. It is evident that stress first rises more quickly before slowing down and eventually leveling out. Before the flow begins, the peak value is linked to static yield stress. **Figure 4b** illustrates the cement paste storage modulus within the linear viscoelastic region (LVR) following the addition of nanoclay. With the addition of nanoclay, the storage modulus G' rises from 2.96×10^4 to $1.72\ 10^5$ Pa [143]. In general, nanomaterials reduce the amount of free water in mixture, which lowers the fluidity or flowability and negatively impacts workability [142].

4.3.2 Nanoclay in modification of polymers

Petroleum-based polymers are not compatible with nanoclay materials due to surface energies differences, requiring surfactants to reduce clay layers' surface energy [45]. Karami et al. [144] studied the impact of microstructure on the rheology of blends of LDPE and polyhydroxy butyrate (PHB) filled with nanoclay and **Figure 5** displays the complicated viscosities of the matching nanocomposites and the neat PHB and LDPE. As shown in **Figure 5**, adding more nanoclay in place of PHB, the complex viscosity would increase significantly. Kolahchi and Kontopoulou [145] found that the neat PHB exhibited a very low viscosity and a Newtonian plateau with minimal shear thinning. The PHB nanocomposite exhibited increased shear thinning behavior with increased addition of Cloisite 10A organoclay, indicating a higher affinity for this material (PHB) [146]. The addition of nanoclay to PHB75/LDPE/N5 or PHB/N5 enhances its elasticity and melt viscosity, thereby improving its processability. Pan et al. [147] found that nanoclay composites significantly improved the resistivity and rheological properties of synthetic-based drilling fluids for high-temperature applications.

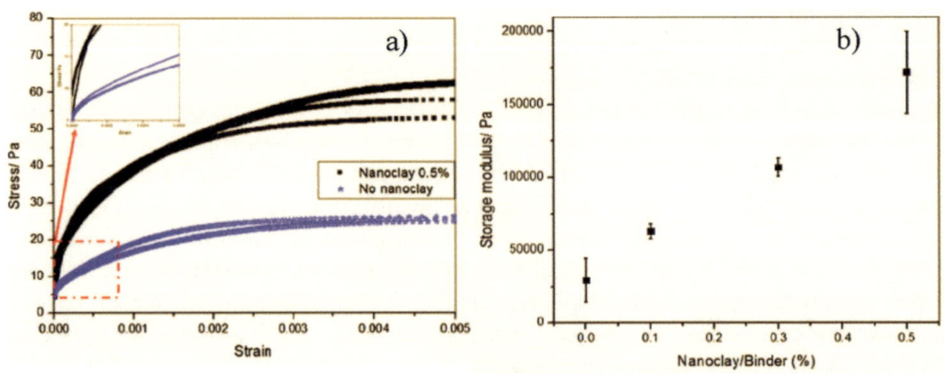

Figure 4.
(a) Stress development of cement pastes without and with 0.5% of nanoclay addition under 0.01 s^{-1} and (b) storage modulus of cement pastes with various nanoclay addition [143].

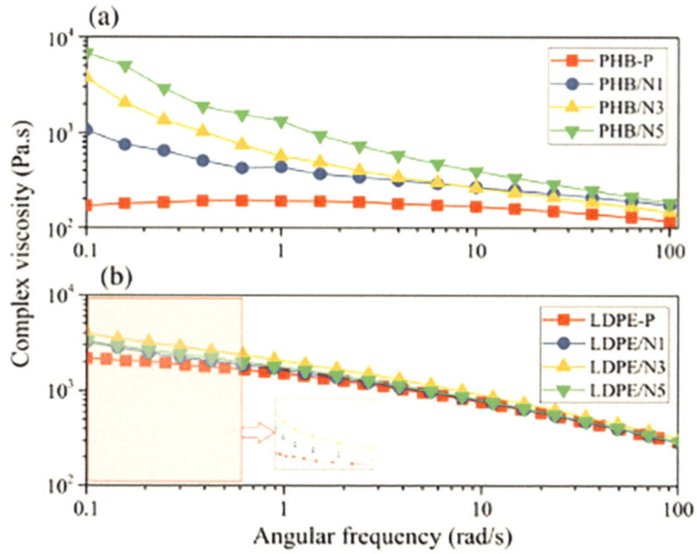

Figure 5.
Complex viscosity as a function of angular frequency for PHB and LDPE nanocomposites (composition, PHB-P, pure, PHB/N1:100/1, PHB/N3:100/3, PHB/N3: 100/5 and similar representation for LDPE) [145].

4.3.3 Nanoclays in modification of asphaltic binder

Nanoclay composites are used in a number of novel applications, such as in asphalt mixtures, where they improve the rutting and fatigue resistance of asphalt mixtures, and enhance the storage stability and the aging resistance of polymer-modified asphalt mixtures [146]. Many research works reported that nanoclays improve the rutting resistance of asphaltic binder [130, 148–152]. Shah and Mir [153] found that combining kaolinite clay and styrene-butadiene-styrene (KC/SBS) in asphalt improves compatibility with SBS, enhancing performance of rutting strength. Amini et al. [148] modified asphalt rubber binder (ARB) with nanoclay to enhance rutting resistance, fatigue properties, and temperature susceptibility. Nanoclay addition improves ARB's performance, reduces viscosity-temperature susceptibility, and temperature susceptibility, while decreasing permanent strain, enhancing rutting resistance, and improving elastic recovery. Another important parameter in asphaltic binder is fatigue resistance which simulates repeated loading cycles of vehicle on pavement road. The role of nanoclay in fatigue resistance of asphaltic binder has been researched extensively [154–157]. Saboo et al. [152] performed the effect of nanoclay on physical and rheological properties of waste cooking oil–modified asphalt binder. The result of the study indicated that addition of nanoclay to WCO-modified asphalt binders improves the physical and rheological properties. The integration of nanotechnology and polymer science can significantly improve the production of modified asphalt mixtures for improved road performance. This alternative has significant potential to improve asphalt surfaces, especially in tropical countries and slow, heavy traffic highways to reduce rutting [150].

4.3.4 Nanoclay in modification of geopolymers

Geopolymer is a three-dimensional polymer having various types of amorphous and semi-crystalline phases with a linkage of Si–O–Al [158]. To produce cellular geopolymer, a stable bubble structure has to be created within an alkali-activated fly ash (AAF) binder paste, which on hardening produces a solid geopolymer matrix and an entrained pore structure [159]. Nanoclays made of montmorillonite and bentonite minerals were used as rheology modifiers. Bentonite typically produces a thickening effect at low concentrations without increasing the elastic response. Montmorillonite has been used as a thixotropic agent in alkali-metal silicates [160]. The result of Gadkar A, Subramaniam KVL [159] confirmed the yield stress of AAF binder pastes is increased with the addition of nanoclay. Montmorillonite is more effective than bentonite in increasing the yield stress of AAF binder paste. A similar increase in the yield stress is obtained at a larger dosage of bentonite.

4.4 Barrier and moisture resistance properties

Nowadays, nanoclays are one of the most widely used nanoparticles in the field of construction and building industries, since they have the potential to enhance the flame retardance, thermal, mechanical, chemical, and barrier properties of building and construction materials [161, 162]. Researchers have investigated the effect of nanoclays on the barrier, chemical and moisture resistance properties of building and construction materials [163–167]. Kim et al. [164] study on organoclay-epoxy nanocomposites revealed that water absorption behavior is influenced by factors like matrix and fiber properties, interfacial adhesion, and voids. The hydrophilic nature of natural fiber-reinforced polymer composites, particularly B/K/Epoxy (11.4%), is the primary driver for water absorption, making it the most water-absorbing composite. **Table 3** illustrates that, out of all the hybrid nanocomposites, B/K/OMMT had the lowest water absorption, with a value of 4.19%. For B/K/MMT and B/K/HNT, the corresponding percentages of water absorption are 4.99 and 5.12%. Because of their hydroxyl groups, natural fibers, especially cellulose and hemicellulose, absorb water [164]. Hybrid nanocomposites can suppress water absorption by adding nanoclay, which reduces void content, creates a moisture barrier, and limits polymer chain motions and relaxation [164, 168]. Ola Bakr Shalby et al.'s [169] study shows that using nanoclay alone as a partial cement replacement reduces water penetration depth from 15 to 9 cm, while the proposed NCSF mix only improves it to 12 cm due to irregular steel fiber shape [169]. Calambas et al. [170] found that adding more than 0.5% MMT to polymeric achira starch/PVA formulations significantly increases the water vapor transmission rate, reducing the materials' barrier capability. The addition of montmorillonite nanoclay to biopolymers-clay nanocomposites altered their water-barrier properties due to their inherent properties. However, sonication of MMT nanoclay is crucial for improving properties, as films with sonicated MMT nanoclay exhibit better wettability and absorbency.

Composites	Maximum moisture content at saturation point, M∞ (%)	Diffusion coefficient, D($\times 10^{-6}$ mm^2/s)
B/K/Epoxy	11.4	1.99
B/K/MMT	4.99	1.26
B/K/HNT	5.12	1.34
B/K/OMMT	4.19	1.19

Table 3.
Maximum moisture content at saturation point and diffusion coefficient of the hybrid composites/nanocomposite [164].

Alimohammad et al. [171] found that adding nanoclay to concrete reduced chloride diffusion coefficient, with the lowest coefficient observed on concrete with 3% nanoclay. After 90 days, the electrical resistivity increased, with 1, 2, and 3% mixes showing a visible decrease in initial current due to lower permeability. Niu et al. [172] found that the addition of nanoclay can improve the concrete's resistance to chloride penetration, as per their observation. The study by [173] found that nanoclay enhances resistance against moisture damage in hot mix asphalt, with 6% of nanoclay showing the most significant improvement in moisture susceptibility (**Figure 6**).

The study by [175] examined the impact of nano-based coatings on concrete under heightened exposures, revealing that the addition of nanoclay significantly impacts the chloride content in the concrete. The addition of nanoclay to reinforcement concrete increased chlorine content from 25 to 49%, enhancing its anticorrosion properties and lowering the rebar corrosion ratio by 7, 30, and 37% after 24 hours with the 1, 3, and 5% nanoclay addition, respectively. Woo et al. [176] found that concrete without coating had the highest chloride content, but coatings like neat silane and nanocomposite significantly reduced it by 92 and 69%, respectively. It was observed by [177] that the chloride diffusion coefficient of cement concrete decreases exponentially with the addition of clay. The study found that adding kaolinite clay to cement concrete reduced the chloride diffusion coefficient to 8.68 and 18.87%, indicating that increasing clay content leads to a reduction in the coefficient.

Li et al. [165] found that organo-montmorillonite nanoclay effectively inhibited bitumen's volatile organic compounds (VOCs) emission, with nanoclay containing octadecyl trimethyl ammonium surfactants showing better inhibition. Umk et al. [178] found that adding nanoclay to water-based coatings increased adhesion and mechanical properties, with 2% being the optimal percentage, despite some adverse effects. The study by [179] found that nanomaterials reduced water absorption ratios and chloride diffusion coefficients in coated concrete. The 0.5% nano-SiO2 polymer-modified cementitious coating showed lower chloride diffusion coefficients, especially at 200 days, indicating higher improvement efficacy in chloride resistance compared to 0.5% nano-TiO2. Abuzeid et al. [167] found that nanoparticles in concrete coatings improve resistance against physical salt attack and salt-frost exposure, but higher nanoclay dosages resulted in inferior performance. Usman et al. [180] studied the bitumen modified with nanoclay/pet(polyethylene terephthalate) blend properties, and noted that penetration decreases with replacement, with optimal values at P0.5% - N1.0% (65.0 mm) and P1.0% - N2.0% (65.4 mm). This is due to, PET

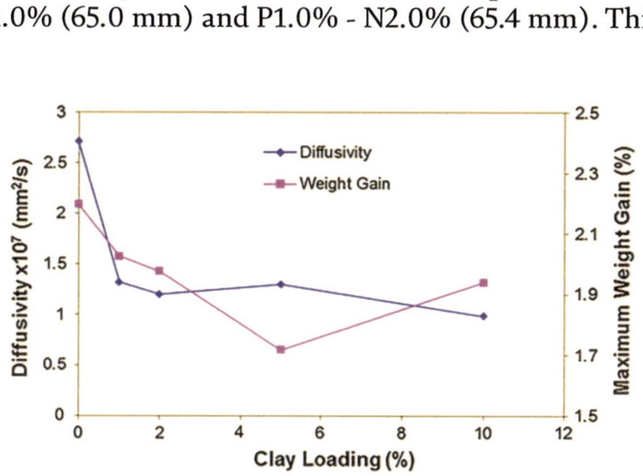

Figure 6.
Variation of diffusivity and maximum weight gain with nanoclay loading. (adapted from Ref. [174]).

absorbing low molecular components of bitumen, causing swelling and movement restriction within asphalt, resulting in reduced penetration of the modified bitumen.

4.5 Fire retardancy

The heating process causes changes in the microstructure of building and construction industry, which in turn affects the structural characteristics of the material [181, 182]. Structures may be efficiently protected by combining passive and active fire prevention systems with management systems like smoke exhaust systems and communication protocols [22, 183]. Researchers are promoting the use of nano-materials as additives to improve thermal resistance in cementitious composites, as they offer superior reactivity compared to micro-scale materials, preventing structural failure [184, 185]. Recently, owing to their smaller particle size than cement grains, nanoclays, such as nano-kaolin, bentonite, mica, and nano-montmorillonite, are utilized as nanofillers in cement-based products which improve chemical, moisture and gas barrier, and mechanical and thermal properties of the material [186]. Nguyen et al. [187] found that the combustion rate (mm/min) of epoxy resin/nanoclay/multiwalled carbon nanotube nanocomposites decreased from 28.41 to 20.80, while the LOI (vol. % O2) increased from 20.6 to 24.6. The addition of nanoclays increased the UL94 from not rated to 18.76 HB. The nanoclay components create a barrier-like substance, slowing down oxygen flow and preventing flammable elements from burning. However, improper dispersion can lead to residual nanoparticles reducing the materials' mechanical properties and fire resistance. Mahesh et al. [188] found that nanoclay addition to vinylester/carbon specimens reduced burning rates by 38 and 46%, respectively, indicating that MMT acts as a good flame retardant, reducing flammability. The improvement in fire behavior is due to the formation of a surface layer during pyrolysis that acts as a heat barrier, preventing heat transfer and increasing surface radiation heat losses [189]. Fillers within the polymer matrix offer a protective clay-rich barrier, thereby reducing the polymer's mass loss rate (MLR) [181, 190–194]. Saraeian et al. [195] investigate the effect of nanoclay particles on the tensile strength and flame retardancy of polystyrene – nanoclay composite. The study showed that adding 5% clay to polystyrene-clay nanocomposite significantly reduced its heat release rate, resulting in a flammability decrease of over 70%, with tests conducted on three nanocomposite groups. Ji & Li [196] found that adding 1.0 wt.% nanoclay to a fire test can reduce mass loss by nearly 13%. The nanoclay morphology transitioned from discontinuous intercalated to well disperse and continued exfoliated, forming insulated layers and a thick charring layer, which partially or completely halted chemical reactions.

Rajaei et al. [197] found that adding 2 wt. % halloysite nano-tube to epoxy slightly reduced the peak heat release rate (PHRR) by about 17%, while incorporating layered double hydroxide (LDH) significantly decreased the PHRR by around 58%. LDH nanoparticles can form a ceramic-like char on the sample's surface, shielding it from heat and flame, but no change in PHRRs was observed with increasing LDH content. The result also shows that adding 18 weight percent APP to EP2LDH and EP2HNT samples reduced PHRRs by 70 and 84%, respectively. Layered nanoclays such as LDH and MMT mitigate the flammability of polymer nanocomposites [198, 199], whilst fibrous nano-additives like carbon nanotube and HNT provide excellent thermal stability which facilitates the formation of chars [199, 200]. Usman et al. [180] studied on properties of bitumen modified with nanoclay/pet (polyethylene terephthalate) blend, and noted that the modified bitumen fire point values were less than the unmodified bitumen (control bitumen). This is due to the strong interfacial interaction between bitumen and nanoclay. Nanoclay shields bitumen from thermo-oxidation due to its "labyrinth" effect, forming a carbonaceous silicate structure that delays volatile product escape and oxygen penetration.

5. Sustainability and environmental considerations

Population growth has caused expansion of cities and, as a result, an increase in the demand for construction and building. On the other hand, excessive use of non-renewable resources, water, and materials, and non-neutrality of emissions are the major challenges facing the sector. This resulted in depletion of natural resources and pollution which are ever-existing problems and increasing at an alarming rate. The building and construction industries are the major consumers of global energy and significant contributors to environmental pollution. It has been identified that the construction and building sector is currently responsible for 40% of solids generation, 36% of global energy demand, 12% of drinking water depletion, 37% of energy-related CO_2 emissions, and 38% of greenhouse gas emissions [201]. Concrete manufacturing, the second most consumed material in the world next to water, is single-handedly responsible for 8% of anthropogenic CO_2 emissions, which results mainly from the Ordinary Portland Cement (OPC) manufacturing used as a binder [202]. Manufacturing of Ordinary Portland Cement (OPC) in particular accounts for 5–8% (roughly 1.35 billion tons of carbon dioxide per year) of greenhouse gas emissions worldwide due to the extensive amount of thermal energy used, which is mainly derived from non-renewable resources such as natural gas, coal, and others [203–208]. Furthermore, research on emissions analysis demonstrated that there is a one-to-one relationship between cement output and CO_2 emissions, which is frightening for the future of human beings [209]. This is further amplified in such a way that it is predicted that the annual global output of cement will reach to 6.1 billion tons by 2050 [210, 211]. Moreover, the use of cement-based concrete as the primary building material is expected to continue to increase to meet the rising need in the construction sector, with a yearly growth rate of 5% in cement production [212, 213]. Continuous repair of damaged buildings from conventional concretes synthesized from OPC, which is labor intensive and expensive process and a concern for environmental mitigation, has also continued to be the challenge for the sector [212]. Therefore, developing more sustainable and environmentally benign concrete mixture options is highly anticipated to lower environmental impacts related to the construction industry. Fortunately, recent developments in green construction are gaining emphasis and are becoming the global agenda for academicians and policymakers [214]. Significant efforts have been put into effect by researchers and industrialists to explore innovative approaches to enhance the performance of construction and building materials and to reduce their environmental impacts. These developments aim to reduce the reliance on cement-based construction in order to minimize carbon dioxide emissions which consequently provide significant environmental benefits. In this regard, geopolymers, ecologically benign cementitious ingredients requiring less energy and producing less CO_2, have emerged as potential alternatives to traditional cement-based construction materials with better performances in compressive strength, acid resistance, creep, and shrinkage [215–219]. Furthermore, the cementitious materials used to produce geopolymers are derived from industrial wastes including slag [220], fly ash [221], and waste glass powder [222]. However, high porosity, poor interfacial bond strength, and delayed strength gain at a later stage in geopolymer-based construction materials, like geopolymer concrete, deter its performance and industrial applications.

Nanoparticles are materials with new insights recently considered outstanding filler materials to alleviate the challenges in geopolymer concrete [223–225]. Geopolymer concrete performance can be improved by incorporating nanoclays, nanoparticles with unique properties, into the mix, and have been considered as a promising solution for the building and construction sector. Environmental and economic benefits of nanoclays, such as their excellent compatibility with water-based systems reducing the need for harmful solvents and volatile organic compounds, and their renewability, have recently created sizable interest in nanoclay-based composites. Research showed that incorporating nanoclays into the geopolymer paste increased the water volume used, consequently lowering the setting time and improving solidity of the paste [226–229], which resulted improvements in self-compaction, ductility, durability, weight reduction, air quality, resistance to chloride penetration, reduction in water and oxygen permeability, lessening in shrinkage, increase in the mechanical strength, and resistance to chemical solvents of the concrete [88, 230]. Optimization in the amount of nanoclays used in the mix determines the best performance; adding a higher percentage may diminish the performance, which is caused by the agglomeration of the nanoparticles [231]. The addition of proper amount of nanoclays to the cementitious mix may lead to synthesis of a new material with improved performances in mechanical properties with significant gains in economic, social, and environmental aspects. As a technical solution to augment properties of concrete, the integration of nanoparticles in concrete mixes has recently promoted and emphasis has been given to the investigation of the effects of the nanoparticles on performance of the composite material.

On the other hand, the use of geopolymer concrete in the construction industry is still in its infant stage. There is no general consensus regarding the influence of nanoparticle additives on the original material, the optimum dosage, or the synergistic effects of the mix components. There is no clear understanding of the mechanism of dispersion, diffusion manner, the hydration process, the chemical reactions at nanoscale, hardening process, overall performance assessment, ease of operation, and others. Particularly, the balance between operational ease and uniform dispersion of nanoparticles presents major challenges in commercialization of nanoclay-reinforced geopolymer concrete in the construction industry. Besides, there have been observations also that the hardening temperature increases with the addition of nanoclays [125]. Additionally, variations in the results of research outputs on the effects of nanoclays on concrete or cement mortar are usually observed due to changes in the type (fly ash, silica fume, etc.) used in the mix and differences in the curing days [201]. Therefore, the behavior of naoclay-reinforced geopolymer concrete is still an active research area for further exploration. Further studies are also needed to develop predictive models to evaluate the correlations between nanoclay particles and the performance enhancements in the geopolymer construction materials; this significantly increases the efficacy of the process and minimizes material wastage.

Despite the gains in environmental impact reduction by using nanoclay-embedded construction materials, the synthesis of the nanoparticles is highly energy-intensive and the process is not yet optimized, which is mainly researched at lab or pilot scale. Inhaling or ingesting high doses of nanoparticles during processing of the mix may cause health effects for construction workers. Hence, studies on the cradle-to-gate life cycle assessment of the nanoparticles showed an increase in other impact categories such as potential ecotoxicity and toxicity [201]. Thus, investigations regarding the environmental impacts of nanoclays addition to mixed designs of construction and building materials are not still sufficient and require further emphasis. Likewise, there are still concerns about the sustainability of concrete structures built from nanoparticle-enabled materials. Environmentally safe and sustainable applications of such materials are demands in prospect for the construction and building industries.

6. Challenges and future perspectives

6.1 Challenges

The absence of consistent dispersion of nanoparticles, which agglomerate in cement, presents difficulties for the building sector [232]. However, the improved hydrophobicity and self-attraction of carbon nanotubes and carbon nanofibers make their application desirable. It is essential to comprehend intricate systems and interface interactions [233]. Ingredients could require further steps, such as purification and functionalization, before they are combined. Nevertheless, this can lead to a weak link forming between the cement and the nanomaterials (carbon nanotubes and carbon nanofibers). Some scientists have created a simple technique to enhance the dispersion of carbon-based nanomaterials in an attempt to tackle this issue.

Since clay-based nanomaterials are hydrophilic, extra caution must be used to regulate the amount of water required in the clay-cement composite. Exchange alteration lowers the hydrophilicity of the interlayer by substituting an organic cation for any calcium or sodium that may have been present. This results in a reduction in water. The main issue to take into account when using nano-coatings on some cultural heritage stones was their photo-induced super hydrophilicity, which could have a negative effect on the stone [234].

Another obstacle to the widespread application of nanoparticles is their compatibility with building materials [235, 236]. In relation to titanium dioxide nanoparticles in particular, some studies have confirmed that there may be more complications when titanium dioxide is added to cement as opposed to coating other substrates like ceramics and glass. Cement has a lower surface area and poor stability [237, 238], which are detrimental to the photocatalytic reaction and have an adverse influence on the use of low stability in building matrices. Its efficiency decreases over time. Generally, regardless of whether it was applied as a coating on the surface or combined with the bulk, the photocatalytic influence seems to diminish after 4 months. One other disadvantage associated with the application of nanoparticles in the construction industry is that, when used to increase strength, these materials also have a larger density, which adds weight to the structure [239].

While nanomaterials (NMs) can enhance the qualities and performance of building materials, there are limitations on their use and commercialization, as well as issues with raw material availability and, as mentioned earlier, high production costs, which make some NMs difficult to obtain [240]. One of the biggest obstacles to using nanomaterials in the building industry is their expensive cost [241]. This is a result of the technology's novelty as well as the sophisticated machinery used in characterization and production procedures. However, costs have been demonstrated to decrease over time, and these costs may decrease much further as fabrication technologies progress. Currently, the higher cost of self-healing concrete compared to traditional OPC-based concretes is mostly caused by their superior attributes, limited knowledge about their application, and lower worldwide manufacturing. Despite their high cost, self-healing concretes based on nanomaterials should be evaluated for adoption due to their long-term benefits. Nanomaterials are anticipated to provide remarkable solutions for resolving any complex problems, leading to their use on a commercial scale and making them affordable [242].

Despite all of this, environmental issues and health risks are two main drawbacks of using nanoparticles in the building industry [243, 244]. Human health may be at risk from construction items with nanotechnology [245–247]. Given that constructive construction is found in a natural setting, all of the materials used in the facilities must, to the greatest extent possible, be environmentally friendly and compatible with the surrounding environment. The most common possible problems include nanoparticles entering groundwater, dispersing through dust into the atmosphere, and being exposed to potentially dangerous materials at the time of construction and during maintenance processes.

6.2 Future perspectives

The results of an extensive analysis of global nanotechnology advancements in construction, along with a survey of construction professionals, indicate that the aforementioned research areas will likely have a significant impact on the built environment and the construction industry in the upcoming 10 years. Notably, given the enormous amounts of materials involved, the most significant influence on the economy and the construction sector during this time is anticipated to come from improvements in production processes and advances in material performance attained through improved understanding and manipulation at the nanoscale [248]. Nanoclay has the potential to become a workable and sustainable solution for improving the performance and sustainability of building materials and practices by solving issues through creative research and development initiatives. In order to fully utilize nanoclay in the construction industry, more study and application are needed in the following areas:

- Refining nanoclay formulations to achieve even higher gains in strength, toughness, and durability

- Creating more environmentally friendly manufacturing processes for nanoclay

- Incorporating nanoclay into eco-friendly building materials that minimize resource consumption and reduce carbon emissions

- Involve the integration of nanoclay additives into 3D printing materials to create structures with enhanced strength, durability, and functionality

- Development of nanoclay additions for creative building materials with features like improved insulation, fire resistance, self-healing, or antibacterial qualities, opening up new opportunities

- Smart nanoclay additives are to be developed that can sense and react to changes in their environment, preventing structural failures and enhancing maintenance procedures

- Involves initiatives to create industry standards for the application of nanoclay in building materials and legal frameworks to handle environmental, health, and safety issues.

7. Conclusion

The utilization of nanoclays as fillers in the building and construction industries holds great potential for enhancing the performance of construction materials. The results from the literature review indicate that the incorporation of nanoclays can lead to improvements in mechanical properties, durability, fire resistance, and thermal insulation. Nanoclays offer unique characteristics, such as high aspect ratios and large specific surface areas, which contribute to enhanced load transfer, interfacial bonding, and barrier properties. However, there are challenges that need to be addressed, including dispersion and compatibility issues, cost considerations, and potential environmental impacts. Future trends in the field of nanoclays in construction include the development of multifunctional materials with additional properties, as well as exploring novel synthesis methods and surface modifications. Further research and development efforts are needed to optimize the manufacturing processes and ensure the sustainability of nanoclay-based materials. The application of nanoclays in the building and construction industries has the potential to revolutionize the field, leading to the development of sustainable, high-performance construction materials.

Author details

Zenamarkos Bantie*, Asmare Tezera, Desalegn Abera and Tesfa Nega
Chemical Engineering Program, Faculty of Chemical and Food, Engineering, Bahir Dar Institute of Technology-Bahir Dar University, Bahir Dar, Ethiopia

*Address all correspondence to: zenamarkosbs@gmail.com

References

[1] Al AT. Clay and clay mineralogy. 2018

[2] Guggenheim S, Ave SO. Report definition of clay and clay mineral: Joint report of the AIPEA nomenclature and CMS nomenclature committees. Clays and Clay Minerals. 1995;**43**:255-256

[3] Belghazdis M, Hachem E. Clay and clay minerals: A detailed review. International Journal of Recent Technology and Applied Science. 2022;**4**:54-75. DOI: 10.36079/lamintang.ijortas-0402.367

[4] Wilson MJ. The Origin and Formation of Clay Minerals in Soils: Past, Present and Future Perspectives. Cambridge, United Kingdom: Cambridge University Press; 1999. pp. 7-25

[5] Cavalcante F. Editorial for Special Issue "Clays, Clay Minerals, and Geology". Switzerland: MDPI Basel; 2021

[6] Woldemariam MH, Belingardi G, Koricho EG, Reda DT. Effects of nanomaterials and particles on mechanical properties and fracture toughness of composite materials: A short review. AIMS Materials Science. 2019;**6**:1191-1212. DOI: 10.3934/matersci.2019.6.1191

[7] Sharma U, Yadav B, Gupta N. Evaluating effects of various mineral admixtures when replaced with a part of cement in concrete. IOP Conference Series: Materials Science and Engineering. IOP Publishing; 2021;**1116**(1):012169

[8] Kushwaha P, Chauhan AS, Swami S, Swami BL. Investigating the effects of nanochemical-based ionic stabilizer and co-polymer on soil properties for pavement construction. International Journal of Geotechnical Engineering. 2019

[9] Christine S, Mana A, Hanafiah MM, Khan AJ. Environmental characteristics of clay and clay- based minerals. Geology, Ecology, and Landscapes. 2017;**9508**:1-7. DOI: 10.1080/24749508.2017.1361128

[10] Repository ZO, Library M. The origin and formation of clay minerals in alpine soils the origin and formation of clay minerals in alpine soils running head: Clay minerals in alpine soils. Hydrogeology, Chemical Weathering, and Soil Formation. 2021:121-137

[11] Attah LE. The Composition and Physical Properties of some Clays of Cross River State, Nigeria. African Journals Online (AJOL). pp. 84-93

[12] Mukherjee S. The science of clays. Netherlands: Springer Science and Business Media. New York: Springer; 2013

[13] Tabatabaie K. Assessment of Nanomaterials Use in Asphalt. 2019;**5**:6-12

[14] Bergaya FBKG, Lagaly G. General introduction: Clays, clay minerals, and clay science. In: Developments in Clay Science. Vol. 5. Elsevier; 2013. pp. 1-19

[15] Nistic R. A Comprehensive Study on the Applications of Clays into Advanced Technologies, with a Particular Attention on Biomedicine and Environmental Remediation. Switzerland: MDPI Basel; 2022

[16] Holtz RD, Kovacs WD. The Characteristics of Clay. Basic A, Specific B the characteristics of clay table of contents. In: Groundwater

& Environmental Engineering. USA: Prentice-Hall; 1981. pp. 1-12

[17] Pusch R. Mechanical properties of clays and clay minerals. In: Handbook of Clay Science. Germany: Science Direct (Elsevier); 2006. pp. 247-260

[18] Gajic IMS, Stojiljkovic ST, Savic IM. Industrial application of clays and clay minerals complimentary contributor copy. Geological Origin, Mechanical Properties and Industrial Applications. 2014:379-402

[19] Velde B. Introduction to clay minerals: Chemistry, origins, uses and environmental significance. London: Chapman & Hall; 1992;**198**

[20] Dwiatmoko MU, Saismana U, Maulanai R. Characteristics of kaolin clay on alluvial formation subdistrict Mataraman based on physical properties and chemical properties. MATEC Web of Conferences. 2019;**280**:03009

[21] Nanocomposites PC. Polymer/Clay Nanocomposites. USA: Michigan State University ProQuest Dissertations Publishing; 2018. DOI: 10.5772/14464

[22] Laím L, Caetano H, Santiago A. Review: Effects of nanoparticles in cementitious construction materials at ambient and high temperatures. Journal of Building Engineering. 2020;**35**(1):102008. DOI: 10.1016/j.jobe.2020.102008

[23] Wagner J. Mechanical Properties of Clays and Clay Minerals. 2nd ed. Elsevier Ltd.; 2013

[24] Theng BKG. The Clay Minerals. 2nd ed. Netherlands: Elsevier B.V; 2012

[25] Valapa RB, Loganathan S, Pugazhenthi G, Thomas S, Varghese TO. An overview of polymer/clay nanocomposites an overview of polymer/clay nanocomposites. Clay-Polymer Nanocomposites; 2017:29-81

[26] Olafusi OS, Sadiku ER, Snyman J, Ndambuki JM, Kupolati WK. Application of nanotechnology in concrete and supplementary cementitious materials : A review for sustainable construction. SN Applied Sciences. 2019;**1**:580

[27] Brightson P, Anand MP, Rajelin DA, Jeparaj P. Application of Nanotechnology in Concrete. 2013;**2**:39-44

[28] Mohajerani A, Burnett L, Smith JV, Kurmus H, Milas J, Arulrajah A, et al. Nanoparticles in Construction Materials and Other Applications, and Implications of Nanoparticle Use. Switzerland: MDPI Basel; 2019. pp. 1-25

[29] Amudhavalli NK. A review on nano technology in concrete. IAEME Publication. 2019;**10**:796-803

[30] Nadu T, Nadu T, Author C. Application of nanoparticles in concrete based on their physical parameters - A review on properties. International Journal of Advanced Research in Engineering and Technology. 2020;**11**: 1537-1549. DOI: 10.34218/IJARET.11.10.2020.147

[31] Mendes TM, Hotza D, Repette WL. Nanoparticles in cement based materials: A review. Reviews on Advanced Materials Science. 2015;**40**:89-96

[32] Jawaid M. Nanoclay reinforced polymer composites. Engineering Materials Series; 2016

[33] Hameed OM, Usman F, Hayder G, Al-ani Y. Investigation of mechanical and thermal performance of nanoclay modified concrete for energy efficiency. Annales de Chimie - Science des Matériaux. 2023;**47**:225-235

[34] Huggett JM. Clay minerals. Encyclopedia of Geochemistry. 2004:358-365. DOI: 10.1016/B0-12-369396-9/00273-2

[35] De Andrade FA, Al-qureshi HA, Hotza D. Measuring and Modeling the plasticity of clays. Materials Research. 2010;**13**:395-399

[36] Fulignati P. Clay minerals in hydrothermal systems. Minerals. 2020;**10**:919

[37] Zhang D, Zhou C, Lin C, Tong D, Yu W. Applied clay science synthesis of clay minerals. Applied Clay Science. 2010;**50**:1-11. DOI: 10.1016/j.clay.2010.06.019

[38] Kumari N, Mohan C. Basics of clay minerals and their characteristic properties. In: Clay and Clay Minerals. USA: Springer; 2021

[39] Saikia P. Clay Nanostructures for Biomedical Applications. Netherlands: Elsevier B.V; 2019

[40] Persano F, Leporatti S. Nano-clays for cancer therapy: State-of-the art and future perspectives. Journal of Personalized Medicine. 2022;**12**:1736

[41] Singh NB. Clays and clay minerals in the construction industry. Minerals. 2022;**12**:301

[42] Zhu PW, Bartos PJM, Porro A. Application of nanotechnology in construction summary of a state-of-the-art report. Materials and Structures. 2004;**37**:649-658

[43] Kinuthia JM. Unfired clay materials and construction. In: Nonconventional and Vernacular Construction Materials. New York: Springer link; 2020

[44] Das BB, Mitra A. Nanomaterials for construction engineering-a review. International Journal of Minerals, Metallurgy and Materials. 2014;**2**:41-46. DOI: 10.7763/IJMMM.2014.V2.96

[45] Majeed K, Jawaid M, Hassan A, Abu Bakar A, Abdul Khalil HPS, Salema AA, et al. Potential materials for food packaging from nanoclay/natural fibres filled hybrid composites. Materials and Design. 2013;**46**:391-410. DOI: 10.1016/j.matdes.2012.10.044

[46] Kumar R, Lokman D, Konala H, Dakuri A, Kirtiraj R. Nano clays and its composites for food packaging applications. International Nano Letters. 2025;**13**:131-153. DOI: 10.1007/s40089-022-00388-8

[47] Rodriguez M, Gomis M, Rico B Nanoclay-based pigments: Synthesis, characterization and application

[48] Sunil BH, Pushpalatha M, Basavaprasad VM, Huvanna TP. The role of nanoclay in the soil stabilization: A short viewpoint. Journal of Geomine. 2023;**1**:92-102. DOI: 10.22077/jgm.2023.6759.1014

[49] Bh S, Pushpalatha M, Vm B, Tp H. Modified nano-clay formulation and their application. International Journal of Chemical Studies. 2018;**6**:705-710

[50] Guo F, Aryana S, Han Y, Jiao Y. A review of the synthesis and applications of polymer-nanoclay composites. Applied Sciences. 2018;**8**:1-29. DOI: 10.3390/app8091696

[51] Yilmaz B, Irmak ET, Turhan Y, Turhan O. Synthesis, characterization and biological properties of intercalated kaolinite nanoclays: Intercalation and biocompatibility. 2019;**19**:83-99. DOI: 10.2478/adms-2019-0007

[52] De Silva RT, Pasbakhsh P, Goh KL, Mishnaevsky L. 3-D computational

model of poly (lactic acid)/halloysite nanocomposites: Predicting elastic properties and stress analysis. Polymer (Guildf). 2014;**55**:6418-6425. DOI: 10.1016/j.polymer.2014.09.057

[53] Jumahat A, Talib AAA, Abdullah A. Wear properties of nanoclay filled epoxy polymers and Fiber reinforced hybrid composites. Engineering Materials. 2016:247-260

[54] Babu, Valapa R, Loganathan S, Pugazhenthi G, Thomas S, Varghese TO. Chapter 2 - An overview of polymer. Clay-Polymer Nanocomposites. 2017:29-81

[55] Okamoto M. Recent advances in polymer/layered silicate nanocomposites: An overview from science to technology. 2006;**22**:756-779. DOI: 10.1179/174328406X101319

[56] Yusoh K, Kumaran SV, Ismail FS. Surface modification of nanoclay for the synthesis of polycaprolactone (PCL) – Clay nanocomposite. MATEC Web of Conferences. 2018;**02005**:1-6

[57] Usmani MA, Khan I, Ahmad N, Bhat AH. Modification of nanoclay systems: An approach to explore various applications. Engineering Materials. 2016:68-75. DOI: 10.1007/978-981-10-1953-1

[58] Yang MS, Huang J, Zhang H, Noël JJ, Hedberg YS, Chen J, et al. Study on the Self-Repairing Effect of Nanoclay in Powder Coatings for Corrosion Protection. New York: Springer link. 2023. pp. 1-22

[59] Jia Z, Dewi T, Wungu K, Agusta K, Gandaryus A. Montmorillonite (MMT) nanoclay in smart coatings for corrosion protection of metal alloy: A brief review. In: Seventh International Conference on Mechanical Engineering Research 2023. Germany: Science Direct (Elsevier); 2024

[60] Wu Q, Liao J, Yang H. Recent advances in kaolinite nanoclay as drug carrier for bioapplications: A review. 2023;**10**:2300672. DOI: 10.1002/advs.202300672

[61] Nazir MS, Haa M, Kassim M. Characteristic properties of nanoclays and characterization of nanoparticulates and nanocomposites. In: Engineering Materials. Germany: Science Direct (Elsevier); 2016. pp. 35-55

[62] Morari G. Clays, clay minerals and ceramic materials based on clay minerals. InTech; 2016

[63] El Amrani I, Hassani E, Sadik C. Geology and mineralogy of clays for nanocomposites: State of knowledge and methodology. In: Clay and Clay Minerals. Cambridge, United Kingdom: Royal Society of Chemistry; 2016

[64] Bergaya F, Maguy Jaber J-FL. Clays and clay minerals. In: Rubber-Clay Nanocomposites: Science, Technology, and Applications. Cambridge, United Kingdom: Green Chemistry; 2011. pp. 3-44

[65] Brigatti MF, Gala E. Structure and mineralogy of clay minerals. In: Handbook of Clay Science. New York: Springer link; 2013

[66] Brigatti MF, Galan E, Theng BKG, Modena I. Structures and mineralogy of clay minerals. In: Handbook of Clay Science. Washington, United States: Amazon Publishing Seattle; 2006

[67] Joussein E, Sciences E, De ÂC, Bag P, North P, Zealand N. Halloysite clay minerals – A review. Clay Minerals. 2005;**40**:383-426. DOI: 10.1180/0009855054040180

[68] Padil V, VT, Kumar A, Murugesan S, Torres-Mendieta R, Wacławek S, Jun

Young Cheong MČ, et al. Sustainable and safer nanoclay composites for multifaceted applications. Green Chemistry. 2022;**24**:3081-3114. DOI: 10.1039/d1gc03949k

[69] Schoonheydt RA, Johnston CT. The surface properties of clay minerals. EMU Notes in Mineralogy. 2011;**11**:37-373

[70] Uddin F. Clays, Nanoclays, and Montmorillonite Minerals. Cambridge, United Kingdom: Cambridge University Press; 2008. DOI: 10.1007/s11661-008-9603-5

[71] Mansi A, Sor NH, Nahla Hilal SMAQ. The impact of nano clay on normal and high- performance concrete characteristics: A review. In: IOP Conf. Series: Earth and Environmental Science. Amsterdam, Netherlands: Elsevier; 2022

[72] Elongation-at-break EB. An overview of polymer–clay nanocomposites. In: Clay-Polymer Nanocomposites. Netherlands: Elsevier Inc; 2017. pp. 29-81

[73] Katti KS, Jasuja H, Jaswandkar SV, Katti DR. Nanoclays in medicine: A new frontier of an ancient medical practice. Materials Advances. 2022;**3**:7484-7500. DOI: 10.1039/D2MA00528J

[74] Mansi A, Sor NH, Nahla Hilal SMAQ. The impact of nano clay on normal and high-performance concrete characteristics: A review. In: Earth and Environmental Science. London, United Kingdom: Taylor & Francis; 2022

[75] Thakur RK, Singh KK, Ramkumar J. Impact of nanoclay filler reinforcement on CFRP composite performance during abrasive water jet machining. Materials and Manufacturing Processes. 2021;**36**:1-10. DOI: 10.1080/10426914.2021.1906896

[76] Beryl JR, Xavier JR. Halloysite for clay – Polymer nanocomposites: Effects of nanofillers on the anti-corrosion, mechanical, microstructure, and flame-retardant properties — A review. Journal of Materials Science. 2023;**58**:10943-10974. DOI: 10.1007/s10853-023-08710-1

[77] Bunea G, Alexa-Stratulat S-M, Mihai P, Ionut OT. Use of clay and titanium dioxide nanoparticles in mortar and concrete—A state-of-the-art analysis. Coatings. 2023;**3**:506

[78] Athulya R, Nandini J, Bhoi TK, Sundararaj R. Recent advances of nanotechnology in wood protection: A comprehensive recent advances of nanotechnology in wood protection: A comprehensive review. Wood Material Science & Engineering. 2023;**19**:1-12. DOI: 10.1080/17480272.2023.2239800

[79] Tan Y, Yang Q, Zheng M, Sarwar MT, Yang H. Multifunctional nanoclay-based hemostatic materials for wound healing: A review. Advanced Healthcare Materials. 2023;**13**:2302700

[80] Mobaraki M, Karnik S, Li Y, Mills DK. Therapeutic applications of halloysite. Applied Sciences. 2022;**12**:87

[81] Peixoto D, Pereira I, Pereira-silva M, Veiga F, Hamblin MR, Lvov Y, et al. Emerging role of nanoclays in cancer research, diagnosis, and therapy. Coordination Chemistry Reviews. 2021;**440**:213956. DOI: 10.1016/j.ccr.2021.213956

[82] Dana K, Sarkar M. Organophilic nature of nanoclay. In: Clay Nanoparticles. Netherlands: INC; 2020. pp. 117-138

[83] Persano F, Gigli G, Leporatti S. Halloysite-based nanosystems for biomedical applications. Clays and Clay Minerals. 2021;**69**:501-521. DOI: 10.1007/s42860-021-00135-8

[84] Perera KY, Hopkins M, Amit K, Jaiswal SJ. Nanoclays - Containing Bio - Based Packaging Materials: Properties, Applications, Safety, and Regulatory Issues. USA: ARROW @ TU Dublin; 2023. DOI: 10.1007/s40097-023-00525-5

[85] Kausar A, Ahmad I, Aldaghri O, Ibnaouf KH, Eisa MH. Nanoclay-reinforced nanocomposite nanofibers—Fundamentals and state-of-the-art developments. Minerals. 2023;**13**(6):817. DOI: 10.3390/min13060817

[86] Jlassi K, Krupa I, Chehimi MM. Overview: Clay preparation, properties, modification. In: Clay-Polymer Nanocomposites. Netherlands: Elsevier Inc; 2017. pp. 1-28

[87] Patel VK, Varia PDJ. Comparative study on effect of nanomaterials As cement replacement on properties of concrete. International Journal for Science and Advance Research in Technology. 2017;**3**:1076-1081

[88] Mohajerani A, Burnett L, Smith JV, Kurmus H, Milas J. Nanoparticles in construction materials and other applications, and implications of nanoparticle use. Switzerland: MDPI Basel; 2019;**12**:1-26. DOI: 10.3390/ma12193052

[89] Chen X. Application of Nano-additives reinforcement towards concrete with excellent mechanical properties: A review. 2022;**9**:5-9. DOI: 10.33552/CTCSE.2022.09.000707

[90] Nadu T. Effects of nanomaterial on. Concrete. 2017;**10**:5-10. DOI: 10.21276/ijee.2017.10.0102

[91] Ali MF, Rashed MT, Bari MA, Razi KM. Effect of zinc oxide nanoparticle on properties of concrete.International Research Journal of Engineering and Technology. 2020;**7**:1026-1029

[92] Reches Y. Nanoparticles as concrete additives: Review and perspectives. Construction and Building Materials. 2018;**175**:483-495. DOI: 10.1016/j.conbuildmat.2018.04.214

[93] Wang W, Li J, Hu J. Unconfined mechanical properties of Nanoclay cement compound modified calcareous Sand of the South China Sea. 2020;**2020**:1-16

[94] Iftikhar S, Mosir P, Mohammad S, Mir S. Potential application of various nanomaterials on the performance of asphalt binders and mixtures: A comprehensive review. International Journal of Pavement Research and Technology. 2023;**16**:1439-1467. DOI: 10.1007/s42947-022-00207-5

[95] Selvasofia SDA, Sarojini E, Moulica G, Thomas S, Tharani M, Saravanakumar PT, et al. Materials today: Proceedings study on the mechanical properties of the nanoconcrete using nano-TiO_2 and nanoclay. Materials Today Proceedings. 2021;**50**:1319-1325. DOI: 10.1016/j.matpr.2021.08.242

[96] Estravís S, Tirado-mediavilla J, Santiago-calvo M, Luis J, Villafañe F, Rodríguez-pérez MÁ. Rigid polyurethane foams with infused nanoclays: Relationship between cellular structure and thermal conductivity. European Polymer Journal. 2016;**80**:1-15. DOI: 10.1016/j.eurpolymj.2016.04.026

[97] Tariq Z, Murtaza M, Mahmoud M. Effects of nanoclay and silica flour on the mechanical properties of class G cement. ACS Omega. 2020;**5**:11643-11654. DOI: 10.1021/acsomega.0c00943

[98] Irshidat MR, Al-saleh MH. Influence of nanoclay on the properties and morphology of cement mortar. KSCE Journal of Civil Engineering.

2018;**22**:4056-4063. DOI: 10.1007/s12205-018-1642-x

[99] Ahmed R, Niyazi M, Kockal U. Nanoparticles used as an ingredient in different types of concrete. SN Applied Sciences. 2021;**3**:1-17. DOI: 10.1007/s42452-021-04461-3

[100] Saba N, Tahir P, Jawaid M. A review on potentiality of nano filler/natural fiber filled polymer hybrid composites. Polymers. 2014;**6**:2247-2273. DOI: 10.3390/polym6082247

[101] Najmi BN, Afina RA, Shakirah SN. Mechanical and morphological properties of nano filler. MATEC Web of Conferences. 2016;**8**:1-5

[102] Šupová M, Martynková GS, Barabaszová K. Effect of nanofillers dispersion in polymer matrices: A review. Science of Advanced Materials. 2014;**3**:1-25. DOI: 10.1166/sam.2011.1136

[103] Ashori A, Nourbakhsh A. Effects of nanoclay as a reinforcement filler on the physical and mechanical. Journal of Composite Materials. 2014;**43**:1869-1875. DOI: 10.1177/0021998309340936

[104] Amjad A, Anjang A, Rahman A, Awais H. A review investigating the influence of nano filler addition on the mechanical, thermal and water absorption properties of cellulosic fibre reinforced polymer composite. Journal of Industrial Textiles. 2022;**51**:65-100. DOI: 10.1177/15280837211057580

[105] Mylsamy B, Kumar S. Impact of nanoclay on mechanical and structural properties of treated Coccinia indica fibre reinforced. Integr Med Res. Journal of Materials Research and Technology. 2019;**8**:1-8. DOI: 10.1016/j.jmrt.2019.09.076

[106] Fu J, Naguib HE. Effect of nanoclay on the mechanical properties of PMMA/clay nanocomposite foams. Journal of Cellular Plastics. 2006;**42**(4):325-342. DOI: 10.1177/0021955X06063517

[107] Khalili SMR, Tavakolian M, Sarabi A. Mechanical properties of nanoclay reinforced epoxy adhesive bonded joints made with composite materials. Journal of Adhesion Science and Technology. 2012;**24**:37-41. DOI: 10.1163/016942410X507650

[108] Mattausch H. Chapter 5. In: Properties and Applications of Nanoclay Composites. Netherlands: Elsevier Inc; 2015

[109] Wang X, Geng T, Han J, Liu C, S hen C, Turng L, et al. Effects of nanoclays on the thermal stability and flame retardancy of microcellular thermoplastic polyurethane nanocomposites. Polymer Composites. 2018;**39**:E1429-E1440. DOI: 10.1002/pc.24340

[110] Collection TS. Nanoparticle-reinforced building materials with applications in civil engineering. Advances in Mechanical Engineering. 2020;**12**:1-4. DOI: 10.1177/1687814020965438

[111] Kaushik S, Sonebi M, Amato G, Perrot A, Kumar U. Influence of nanoclay on the fresh and rheological behaviour of 3D printing mortar materials today: Proceedings influence of nanoclay on the fresh and rheological behaviour of 3D printing mortar. Materials Today Proceedings. 2022. DOI: 10.1016/j.matpr.2022.01.108

[112] Rafiee R, Shahzadi R. Mechanical properties of nanoclay and nanoclay reinforced polymers: A review. Polymer Composites. 2019;**40**(2):431-445

[113] Qi B. Investigation of the mechanical properties of DGEBA-based epoxy resin with nanoclay additives.

2006;**75**:514-519. DOI: 10.1016/j.compstruct.2006.04.032

[114] Ho M, Lam C, Lau K, Ng DHL, Hui D. Mechanical properties of epoxy-based composites using nanoclays. 2006;**75**:415-421. DOI: 10.1016/j.compstruct.2006.04.051

[115] Nguyen TA, Mai T, Pham H, Dang TH, Do TH, Nguyen QT. Study on mechanical properties and fire resistance of epoxy nanocomposite reinforced with environmentally friendly additive: Nanoclay I. 2020;**30E**:2020

[116] Shettar M, Kini UA, Sharma SS, Hiremath P. ScienceDirect study on mechanical characteristics of Nanoclay reinforced polymer composites. Materials Today Proceedings. 2017;**4**:11158-11162. DOI: 10.1016/j.matpr.2017.08.081

[117] Padil VVT, Kumar KPA, Murugesan S. Sustainable and safer nanoclay composites for. 3081-3114. 2022. DOI: 10.1039/d1gc03949k

[118] Kundie F, Azhari CH, Muchtar A, Ahmad ZA. Effects of filler size on the mechanical properties of polymer-filled dental composites: A review of recent developments. 2018;**29**:141-165

[119] Ozsoy I. The influence of micro- and nano-filler content on the mechanical properties of epoxy composites. 2015;**61**:601-609. DOI: 10.5545/sv-jme.2015.2632

[120] Farzadnia N, Abdullah A, Ali A, Demirboga R, Parvez M. Cement and concrete research effect of halloysite nanoclay on mechanical properties, thermal behavior and microstructure of cement mortars. Cement and Concrete Research. 2013;**48**:97-104. DOI: 10.1016/j.cemconres.2013.03.005

[121] Saleh AN, Attar AA, Ahmed OK, Mustafa SS. Improving the thermal insulation and mechanical properties of concrete using Nano-SiO2. Results in Engineering. 2021;**12**:100303. DOI: 10.1016/j.rineng.2021.100303

[122] Hakamy A, Shaikh FUA, Low IM. Thermal and mechanical properties of hemp fabric-reinforced nanoclay-cement nanocomposites. Journal of Materials Science. 2014;**49**:1684-1694. DOI: 10.1007/s10853-013-7853-0

[123] Irshidat MR, Al-saleh MH. Thermal performance and fire resistance of nanoclay modified cementitious materials. Construction and Building Materials. 2018;**159**:213-219. DOI: 10.1016/j.conbuildmat.2017.10.127

[124] Hamed N, El-feky MS, Kohail M, Nasr EAR. Effect of nano-clay de-agglomeration on mechanical properties of concrete. Construction and Building Materials. 2019;**205**:245-256. DOI: 10.1016/j.conbuildmat.2019.02.018

[125] Hakamy A, Shaikh FUA, Low IM. Characteristics of nanoclay and calcined nanoclay-cement nanocomposites. Composites: Part B. 2015;**78**:174-184. DOI: 10.1016/j.compositesb.2015.03.074

[126] Fan Y, Zhang S, Wang Q, Shah SP. The effects of nano-calcined kaolinite clay on cement mortar exposed to acid deposits. 2016;**102**:486-495. DOI: 10.1016/j.conbuildmat.2015.11.016

[127] Shen J, Liang J, Lin X, Lin H, Yu J. Recent progress in polymer-based building materials. 2020;**2020**

[128] Sepet H, Tarakcioglu N, Misra RDK. Investigation of mechanical, thermal and surface properties of nanoclay/HDPE nanocomposites produced industrially by melt mixing approach. 2015. DOI: 10.1177/0021998315615653

[129] Akelah A, Rehab A, Abdelwahab M, Betiha MA. Synthesis and thermal

properties of nanocomposites based on exfoliated organoclay polystyrene and poly(methylmethacrylate). Nano. 2017;**3**:20-29. DOI: 10.1080/20550324.2017.1316599

[130] Rusbintardjo G, Rosli M, Izzi N. Fundamental and rheological properties of oil palm fruit ash modified bitumen. Construction and Building Materials. 2013;**49**:702-711. DOI: 10.1016/j.conbuildmat.2013.08.056

[131] Mersha DA, Sendekie ZB. High-temperature performance enhancement of bitumen by waste PET-derived polyurethane. 2022;**2022**

[132] Liu G, Fang S, Wang Y, Liu J, Liang Y, Cao T, et al. Emission of volatile organic compounds in crumb rubber modified bitumen and its inhibition by using montmorillonite Nanoclay. Polymers (Basel). 2023;**15**. DOI: 10.3390/polym15061513

[133] Gardênia L, Farias AT, Leitinho JL, Amoni BDC, Bastos JBS, Soares JB, et al. Effects of nanoclay and nanocomposites on bitumen rheological properties. 2016;**125**:873-883. DOI: 10.1016/j.conbuildmat.2016.08.127

[134] Ghoreishi A, Koosha M, Nasirizadeh N. Modification of bitumen by EPDM blended with hybrid nanoparticles: Physical, thermal, and rheological properties. Journal of Thermoplastic Composite Materials. 2020;**33**:343-356. DOI: 10.1177/0892705718805536

[135] Yilmaz B, Münir A, Gürbüz HE. Assessment of thermal properties of Nanoclay-modified bitumen. Arabian Journal for Science and Engineering. 2023;**48**:4595-4607. DOI: 10.1007/s13369-022-07142-4

[136] Assaedi H, Shaikh FUA, Low IM. Effect of nano-clay on mechanical and thermal properties of geopolymer. Integr Med Res. Journal of Asian Ceramic Societies. 2015:1-10. DOI: 10.1016/j.2015.10.004

[137] Khater H. Effect of nano-clay on alkali activated water-cooled slag geopolymer. British Journal of Applied Science & Technology. 2013;**3**:764-776. DOI: 10.9734/bjast/2014/2690

[138] Nazar S, Yang J, Thomas BS, Azim I, Kashif S, Rehman U. Rheological properties of cementitious composites with and without nano-materials: A comprehensive review. Journal of Cleaner Production. 2020;**272**:122701. DOI: 10.1016/j.jclepro.2020.122701

[139] Kawashima S, Wang K, Douglas R, Hong J, Tregger N. Cement and concrete research a review of the effect of nanoclays on the fresh and hardened properties of cement-based materials. Cement and Concrete Research. 2021;**147**:106502. DOI: 10.1016/j.cemconres.2021.106502

[140] Dejaeghere I, Sonebi M, De SG. Influence of nano-clay on rheology, fresh properties, heat of hydration and strength of cement-based mortars. Construction and Building Materials. 2019;**222**:73-85. DOI: 10.1016/j.conbuildmat.2019.06.111

[141] Kawashima S, Hou P, Corr DJ, Shah SP. Modification of cement-based materials with nanoparticles. Cement and Concrete Composites. 2013;**36**:8-15. DOI: 10.1016/j.cemconcomp.2012.06.012

[142] Varela H, Barluenga G, Palomar I. Influence of nanoclays on flowability and rheology of SCC pastes. Construction and Building Materials. 2020;**243**:118285. DOI: 10.1016/j.conbuildmat.2020.118285

[143] Qian Y, Kawashima S. Cement and concrete research use of creep recovery

protocol to measure static yield stress and structural rebuilding of fresh cement pastes. Cement and Concrete Research. 2016;**90**:73-79. DOI: 10.1016/j.cemconres.2016.09.005

[144] Karami S, Nazockdast H, Ahmadi Z, Rabolt JF, Noda I, Chase DB. Microstructure effects on the rheology of Nanoclay-filled PHB /LDPE blends. 2019. DOI: 10.1002/pc.25273

[145] Kolahchi AR, Kontopoulou M. Chain extended poly(3-hydroxybutyrate) with improved rheological properties and thermal stability, through reactive modification in the melt state. Polymer Degradation and Stability. 2015;**121**:222-229. DOI: 10.1016/j.polymdegradstab.2015.09.008

[146] Yang J, Tighe S. A review of advances of nanotechnology in asphalt mixtures. 2013;**96**:1269-1276. DOI: 10.1016/j.sbspro.2013.08.144

[147] Pan D, Vipulanandan C, Amani N, Reddy SA, Chockalingam CG. OTC-28751-MS Effects of Nanoclay on the Rheological Properties and Resistivity of Synthetic Based Drilling Fluids under High Temperature. 2018

[148] Amini A, Ziari H, Amir S, Sadat N, Goli A. Rutting resistance, fatigue properties and temperature susceptibility of nano clay modified asphalt rubber binder. Construction and Building Materials. 2020:120946. DOI: 10.1016/j.conbuildmat.2020.120946

[149] Alireza S, Tanzadeh J, Mirsepahi M. Laboratory evaluation of the composition of nano-clay , nano-lime and SBS modifiers on rutting resistance of asphalt binder. Construction and Building Materials. 2020;**238**:117592. DOI: 10.1016/j.conbuildmat.2019.117592

[150] Carlesso GC, Trichês G, de Melo JVS, Marcon MF, Thives LP, da Luz LC. Evaluation of rheological behavior, resistance to permanent deformation, and resistance to fatigue of asphalt mixtures modified with nanoclay and SBS polymer. Applied Sciences. 2019;**9**. DOI: 10.3390/app9132697

[151] Ameri M, Nobakht S, Bemana K, Vamegh M. Effects of nanoclay on hot mix asphalt performance. 2016. DOI: 10.1080/10916466.2016.1164714

[152] Saboo N, Sukhija M, Singh G. Effect of Nanoclay on physical and rheological properties of waste cooking oil–modified asphalt binder. Journal of Materials in Civil Engineering. 2021;**33**:1-13. DOI: 10.1061/(asce)mt.1943-5533.0003598

[153] Mosir P, Mohammad S, Mir S. Effect of kaolinite clay/SBS on rheological performance of asphalt binder. Innovative Infrastructure Solutions. 2020;**5**:1-14. DOI: 10.1007/s41062-020-0270-8

[154] Johnson TW. Application of Nanoclay Materials in Asphalt Pavements. pp. 1-16

[155] Harraz HZ. Nano clay and its applications. Apple Minerals. 2016;**10**

[156] Ramakrishnan S, Krishnamurthy K, Rajasekar R, Rajeshkumar G. An experimental study on the effect of nano-clay addition on mechanical and water absorption behaviour of jute fibre reinforced epoxy composites. 2019. DOI: 10.1177/1528083718792915

[157] Jun LY, Khan FSA, Mubarak NM, Yon LS, Bing CH, Khalid M, Abdullah EC. Importance of nanomaterials in engineering application. 2021

[158] Awwad FA, Ismail EAA, Noor S, Awwad FA. Synthesis and

characterization of novel iron-modified geopolymer cement from laterite clay as low energy material. 2024. DOI: 10.1063/5.0177022

[159] Gadkar A, Subramaniam KVL. Rheology control of alkali-activated fly ash with nano clay for cellular geopolymer application. Construction and Building Materials. 2021;**283**:122687. DOI: 10.1016/j.conbuildmat.2021.122687

[160] Cheng DC. Thixotropy. International Journal of Cosmetic Science. 1987;**9**:151-191. DOI: 10.1111/j.1467-2494.1987.tb00472.x

[161] Beryl JR, Xavier JR. Mechanical and corrosion protection properties of polymer-clay nanocomposite coatings for mild steel in marine environment. 2020

[162] Rafiq A, Merah N. Nanoclay enhancement of flexural properties and water uptake resistance of glass fiber-reinforced epoxy composites at different temperatures. 2018. DOI: 10.1177/0021998318781220

[163] Kim J, Hu C, Woo RSC, Sham M. Moisture barrier characteristics of organoclay – Epoxy nanocomposites. Composites Science and Technology. 2005;**65**:805-813. DOI: 10.1016/j.compscitech.2004.10.014

[164] Sand S, Jawaid M, Sultan MTH, Alothman OY. Effects of nanoclay on physical and dimensional stability of bamboo/Kenaf/nanoclay reinforced epoxy hybrid nanocomposites. Integr Med Res. 2020. DOI: 10.1016/j.jmrt.2020.03.114

[165] Li L, Wu S, Liu G, Cao T, Amirkhanian S. Effect of organo-montmorillonite nanoclay on VOCs inhibition of bitumen (a) Octadecyl trimethyl ammonium cation (b) benzyl dimethyl hexadecyl ammonium cation. Construction and Building Materials. 2017;**146**:429-435. DOI: 10.1016/j.conbuildmat.2017.04.040

[166] Cui Y, Kumar S, Kona R, Van HD. Gas barrier properties of polymer/clay nanocomposites. RSC Advances. 2015;**5**:63669-63690. DOI: 10.1039/C5RA10333A

[167] Abuzeid MA, Bassuoni MT, Sakr MR. Effect of polymer/nano-clay coatings on the performance of concrete with high-content supplementary cementitious materials under harsh exposures. 2024

[168] Jesuarockiam Naveen MJ, Zainudin ES, Yahaya MTHS, Yahaya R. Improved mechanical and moisture-resistant properties of woven hybrid epoxy composites by graphene nanoplatelets (GNP). Materials (Basel). 2019

[169] Shalby OB, Elkady HM, Nasr EAR, Kohail M. Assessment of mechanical and fire resistance for hybrid nano-clay and steel fibres concrete at different curing ages. Mechanical and Fire Resistance. 2019;**11**:189-203. DOI: 10.1108/JSFE-06-2019-0024

[170] Calambas HL, Fonseca A, Adames D, Aguirre-loredo Y. Physical-mechanical behavior and water-barrier properties of biopolymers-clay nanocomposites. 2021

[171] Alimohammad M, Langaroudi M, Mohammadi Y. Effect of nano-clay on workability, mechanical, and durability properties of self-consolidating concrete containing mineral admixtures. Construction and Building Materials. 2018;**191**:619-634. DOI: 10.1016/j.conbuildmat.2018.10.044

[172] Niu X, Li Q, Hu Y, Tan Y, Liu C. Properties of cement-based materials

incorporating nano-clay and calcined nano-clay: A review. 2021;**284**. DOI: 10.1016/j.conbuildmat.2021.122820

[173] Ameri M, Nobakht S, Bemana K, Vamegh M. Effects of nanoclay on hot mix asphalt performance. 2016:6466. DOI: 10.1080/10916466.2016.1164714

[174] Manmzm G. Mechanical Properties and Water Uptake of Epoxy – Clay Nanocomposites Containing Different Clay Loadings. 2013. pp. 3798-3804. DOI: 10.1007/s10853-013-7180-5

[175] Sakr MR, Bassuoni MT. Effect of Nano-Based Coatings on Concrete under Aggravated Exposures. 322020. DOI: 10.1061/(ASCE)MT.1943-5533.0003349

[176] Woo RSC, Zhu H, Chow MMK, Leung CKY, Kim J. Barrier Performance of Silane-Clay Nanocomposite Coatings on Concrete Structure. Compos Sci Technol; 2008. pp. 1-22

[177] Fan Y, Zhang S, Kawashima S, Shah SP. Cement & Concrete Composites Influence of kaolinite clay on the chloride diffusion property of cement-based materials. Cement and Concrete Composites. 2014;**45**:117-124. DOI: 10.1016/j.cemconcomp.2013.09.021

[178] Umk A, Tumirah K, Hamdan H, Sh L. Effects of nanoclay contents on the properties of water-based coating. Journal of Tropical Forest Science. 2019;**31**:353-361

[179] Li G, Ding Y, Gao T, Qin Y, Lv Y, Wang K. Chloride resistance of concrete containing nanoparticle-modified polymer cementitious coatings. Construction and Building Materials. 2021;**299**:123736. DOI: 10.1016/j.conbuildmat.2021.123736

[180] Usman MI, Murana AA, Kaura JM, Ochepo J. Properties of bitumen modified with nanoclay/pet (polyethylene terephthalate). Blend. 2023;**2657**:192-203. DOI: 10.36348/sjce.2023.v07i08.003

[181] Ahmed W, Baloch WL, Khushnood RA. Fire performance of concrete containing nano-fibers and graphite nano-particles. In: Smart Nanoconcretes and Cement-Based Materials. Elsevier Inc.; 2020. pp. 297-311

[182] Vakhitova LN. Fire retardant nanocoating for wood protection. In: Long-Term Performance & Durability of Masonry Structures. Elsevier Ltd; 2019. pp. 361-391

[183] Mckenna ST, Jones N, Peck G, Dickens K, Pawelec W, Oradei S, et al. Fire behaviour of modern façade materials – Understanding the Grenfell tower fire. Journal of Hazardous Materials. 2019;**368**:115-123. DOI: 10.1016/j.jhazmat.2018.12.077

[184] Nguyen Q, Ngo T, Moinuddin K, Mendis P. Use of nanoclay to improve fire performance of glass fibre composites. Proceedings of SPIE - The International Society for Optical Engineering. Victoria Univ., building 4, hoppers lane. 2013;**8793**:1-7. DOI: 10.1117/12.2027959

[185] He W, Song P, Yu B, Fang Z, Wang H. Flame retardant polymeric nanocomposites through the combination of nanomaterials and conventional flame retardants. Progress in Materials Science. 2020;**114**:100687. DOI: 10.1016/j.pmatsci.2020.100687

[186] Necolau I, Damian CM, Claudiu R, Chiriac A, Mihail G, Vasile E, et al. Layered clay – graphene oxide nanohybrids for the reinforcement and fire-retardant properties of polyurea matrix. Polymers. 2022;**14**:66

[187] Nguyen TA, Nguyen QT, Bach TP. Mechanical properties and flame

retardancy of epoxy resin/nanoclay/multiwalled carbon nanotube nanocomposites. 2019;**2019**

[188] Zhang J, Hereid J, Hagen M, Bakirtzis D, Delichatsios MA, Fina A, et al. Effects of nanoclay and fire retardants on fire retardancy of a polymer blend of EVA and LDPE. Fire Saf J J. 2009;**44**:504-513. DOI: 10.1016/j.firesaf.2008.10.005

[189] Mahesh KRV, Murthy HNN, Swamy BEK, Sharma SC. Mechanical, thermal and fire retardation behaviours of nanoclay/vinylester nanocomposites. 2011;**5**:401-411. DOI: 10.1007/s11706-011-0149-x

[190] Abdalla JA, Skariah B, Hawileh RA, Yang J, Bhushan B. Influence of nano-TiO2, nano-Fe2O3, nanoclay and nano-CaCO3 on the properties of cement/geopolymer concrete. Cleaner Materials. 2022;**4**. DOI: 10.1016/j.clema.2022.100061

[191] Kim Y, Lee S, Yoon H. Fire-Safe Polymer Composites: Flame-Retardant Effect of Nanofillers. Polymers. 2021;**13**:540. DOI: 10.3390/polym13040540

[192] Nguyen TA. Study on the synergies of Nanoclay and MWCNTs to the flame retardant and mechanical properties of epoxy nanocomposites. 2021;**2021**:1-8

[193] Pachta V, Tsardaka E, Stefanidou M. The role of flame retardants in cement mortars exposed at elevated temperatures. Construction and Building Materials. 2021;**273**:122029. DOI: 10.1016/j.conbuildmat.2020.122029

[194] Wu Y, Zhang Y, Ju J, Yan H, Huang X, Tan Y. Advances in halloysite nanotubes. Polysaccharide Nanocomposite Preparation and Applications. 2019;**5**

[195] Saraeian P, Ansari K, Tavakoli HR, Soltani E. The effect of nanoclay particles on the tensile strength and flame Retardancy of polystyrene – Nanoclay composite. 2012;**2**:12330-12336

[196] Ji G, Li G. Effects of nanoclay morphology on the mechanical , thermal, and fire-retardant properties of vinyl ester based nanocomposite. 2008;**498**:327-334. DOI: 10.1016/j.msea.2008.08.003

[197] Rajaei M, Kim NK, Bickerton S, Bhattacharyya D. A comparative study on effects of natural and synthesised nano-clays on the fi re and mechanical properties of epoxy composites. Composites: Part B. 2019;**165**:65-74. DOI: 10.1016/j.compositesb.2018.11.089

[198] Kiliaris P, Papaspyrides CD. Polymer/layered silicate (clay) nanocomposites: An overview of flame retardancy. Progress in Polymer Science. 2010;**35**:902-958. DOI: 10.1016/j.progpolymsci.2010.03.001

[199] Bar M, Alagirusamy R, Das A. Flame retardant polymer composites. Fibers and Polymers. 2015;**16**:705-717. DOI: 10.1007/s12221-015-0705-6

[200] Laoutid F, Bonnaud L, Alexandre M. New prospects in flame retardant polymer materials: From fundamentals to nanocomposites. 2009

[201] Maria De Los Ángeles Ortega Del Rosario Del O, Beermann K. Environmentally responsive materials for building envelopes: A review on manufacturing and biomimicry-based approaches. 2023

[202] Webb M. Biomimetic building facades demonstrate potential to reduce energy consumption for different building typologies in different climate zones. Clean Technologies and

Environmental Policy. 2022;**24**:493-518. DOI: 10.1007/s10098-021-02183-z

[203] Kathirvel P. Effect of replacement of slag on the mechanical properties of flyash based Geopolymer concrete. 2015:3-8

[204] Amritphale SS, Mishra D, Mudgal M, Chouhan RK, Chandra N. A novel green approach for making hybrid inorganic- organic geopolymeric cementitious material utilizing fly ash and rice husk. Biochemical Pharmacology. 2016;**4**:3856-3865. DOI: 10.1016/j.jece.2016.08.015

[205] Kishore K, Gupta N. Mechanical characterization and assessment of composite geopolymer concrete. Materials Today Proceedings. 2021. DOI: 10.1016/j.matpr.2020.06.319

[206] Malkawi AB. Effect of aggregate on the performance of fly-ash-based geopolymer concrete. 2023

[207] Özcan A. Evaluation of sulfate and salt resistance of ferrochrome slag and blast furnace slag-based geopolymer concretes. 2019:1607-1621. DOI: 10.1002/suco.201900061

[208] Hassan A, Arif M, Shariq M. Use of geopolymer concrete for a cleaner and sustainable environment e a review of mechanical properties and microstructure. Journal of Cleaner Production. 2019;**223**:704-728. DOI: 10.1016/j.jclepro.2019.03.051

[209] Gunasekara C, Law DW, Setunge S, Sanjayan JG, Piper M, Piper M. Zeta potential, gel formation and compressive strength of low calcium fly ash geopolymers. Construction and Building Materials. 2015;**95**:592-599. DOI: 10.1016/j.conbuildmat.2015.07.175

[210] Ranjbar N, Zhang M. Fiber-reinforced geopolymer composites: A review. Cement and Concrete Composites. 2020;**107**:103498. DOI: 10.1016/j.cemconcomp.2019.103498

[211] Camargo MM, Taye EA, Roether JA, Redda DT, Boccaccini AR. A review on natural fiber-reinforced geopolymer and cement-based composites. Materials. 2020:1-29

[212] Jitha TKJPT, Pattaje SK. Studies on the strength development of lime – Pozzolanacement – soil – brick powder based geopolymer composites. Journal of The Institution of Engineers (India): Series A. 2019;**100**:329-336. DOI: 10.1007/s40030-018-0350-3

[213] Sarker PK, Haque R, Ramgolam KV. Fracture behaviour of heat cured fly ash based geopolymer concrete. Journal of Materials. 2013;**44**:580-586. DOI: 10.1016/j.matdes.2012.08.005

[214] Vyas PCM, Bhatt PDR. Concept of green concrete using construction demolished waste as recycled coarse aggregate. International Journal of Engineering Trends and Technology (IJETT). 2013;**4**:3160-3165. ISSN: 2231-5381 ISSN: 2231-5381

[215] Gupta N, Kishore K, Kumar K, Joshi TC. Influence of industrial by-products on the behavior of geopolymer concrete for sustainable development. Indian Journal of Engineering and Materials Science. 2021;**28**:433-445

[216] Lavanya G, Jegan J. Durability study on high calcium fly ash based geopolymer concrete. Advances in Materials Science and Engineering. 2015;**2015**

[217] Davidovits J, France S. Geopolymer chemistry and sustainable development Geopolymer chemistry and sustainable development. The poly(sialate) term inology: A very useful and simple model

for the promotion and understanding of green-chemistry. 2018

[218] Wang K, Lemougna PN, Tang Q, Li W, He Y, Cui X. Low temperature depolymerization and polycondensation of a slag-based inorganic polymer. Ceramics International. 2017;**43**(12):9067-9076. DOI: 10.1016/j.ceramint.2017.04.052

[219] Nguyen TT, Goodier CI, Austin SA. Factors affecting the slump and strength development of geopolymer concrete. Construction and Building Materials. 2020;**261**:119945. DOI: 10.1016/j.conbuildmat.2020.119945

[220] Duxson P. Geopolymer precursor design. In: Geopolymers. Woodhead Publishing Limited; 2004. pp. 37-49

[221] Krishnarao S, Sravana P. Experimental investigation on pozzolanic effect of fly ash in roller compacted concrete pavement using manufactured sand as fine aggregate. International Journal of Applied Engineering Research. 2016

[222] Vasudevan G, Ganis S. Performance of using waste glass powder in concrete As replacement of cement. American Journal of Engineering Research (AJER). 2013:175-181

[223] Khater HM. Effect of Calcium on Geopolymerization of Aluminosilicate Wastes. 2012. pp. 92-101. DOI: 10.1061/(ASCE)MT.1943-5533.0000352

[224] Thomas BS. The effects of nano- and micro-particle additives on the durability and mechanical properties of mortars exposed to internal and external sulfate attacks. Results in Physics. 2017;**7**:843-851. DOI: 10.1016/j.rinp.2017.02.009

[225] Sobolev K, Gutiérrez MF. How nanotechnology can change the concrete world. In: Concrete Nanotechnology. 2005

[226] Taylor P, Gonzalez M, Cao J, Tighe SL. State-of-the-art report on use of nano-materials in concrete. International Journal of Pavement Engineering:37-41. DOI: 10.1080/10298436.2014.893327

[227] Nasution A, Imran I, Abdullah M. Improvement of concrete durability by nanomaterials. Procedia Engineering. 2015;**125**:608-612. DOI: 10.1016/j.proeng.2015.11.078

[228] Birgisson B, Taylor P, Armaghani J, Shah SP. American road map for research for nanotechnology-based concrete materials. 2007. DOI: 10.3141/2142-20

[229] Hanus MJ, Harris AT. Progress in materials science nanotechnology innovations for the construction industry. Progress in Materials Science. 2013;**58**:1056-1102. DOI: 10.1016/j.pmatsci.2013.04.001

[230] Alireza S, Mojtaba S. Effect of bubble based degradation on the physical properties of Single Wall carbon nanotube/epoxy resin composite and new approach in bubbles reduction. Composites: Part A. 2016;**90**:457-469. DOI: 10.1016/j.compositesa.2016.08.015

[231] Li H, Xiao H, Yuan J, Ou J. Microstructure of cement mortar with nano-particles. 2004;**35**:185-189. DOI: 10.1016/S1359-8368(03)00052-0

[232] Papanikolaou I, Ribeiro de Souza L, Litina C, Al-Tabbaa A. Investigation of the dispersion of multi-layer graphene nanoplatelets in cement composites using different superplasticiser treatments. Construction and Building Materials. 2021;**293**:123543. DOI: 10.1016/j.conbuildmat.2021.123543

[233] Sanchez F, Sobolev K. Nanotechnology in concrete - a review. Construction and Building Materials. 2010;**24**:2060-2071. DOI: 10.1016/j.conbuildmat.2010.03.014

[234] Colangiuli D, Calia A, Bianco N. Novel multifunctional coatings with photocatalytic and hydrophobic properties for the preservation of the stone building heritage. Construction and Building Materials. 2015;**93**:189-196. DOI: 10.1016/j.conbuildmat.2015.05.100

[235] Wu S, Tahri O. State-of-art carbon and graphene family nanomaterials for asphalt modification. Road Materials and Pavement Design. 2021;**22**:735-756

[236] Ali RA, Kharofa OH. The impact of nanomaterials on sustainable architectural applications smart concrete as a model. Materials Today Proceedings. 2021;**42**:3010-3017

[237] Feng H, Zhao X, Li L, Zhao X, Gao D. Water stability of bonding properties between nano-Fe2O3-modified magnesiumphosphate-cement mortar and steel fibre. Construction and Building Materials. 2021;**291**:123316

[238] Hou L, Li J, Lu Z, Niu Y. Influence of foaming agent on cement and foam concrete. Construction and Building Materials. 2021;**280**:122399

[239] Srinivas K. Nanomaterials for concrete technology. International Journal of Civil, Structural, Environmental and Infrastructure Engineering Research. 2014;**1**:79-90

[240] Saleem H, Zaidi SJ, Alnuaimi NA. Recent advancements in the nanomaterial application in concrete and its ecological impact. Materials. 2021;**14**:6387

[241] Onaizi AM, Huseien GF, Lim NHAS, Amran M, Samadi M. Effect of nanomaterials inclusion on sustainability of cement-based concretes: A comprehensive review. Construction and Building Materials. 2021;**306**:124850.271

[242] Rana AK, Rana SB, Kumari A, Kiran V. Significance of nanotechnology in construction engineering. International Journal of Recent Trends in Engineering. 2009;**1**:46

[243] Mallakpour S, Hussain CM, Ajith S, Arumugaprabu V. Environmental and occupational health hazards of nanomaterials in construction sites. In: Handbook of Consumer Nanoproducts. Springer; 2021. pp. 1-12

[244] Onyango J. Health impacts of building materials on construction workers. In: Ecological and Health Effects of Building Materials. Cham, Switzerland: Springer; 2022. pp. 543-566

[245] Santhosh G, Nayaka GP. Nanoparticles in construction industry and their toxicity. In: Ecological and Health Effects of Building Materials. Cham, Switzerland: Springer; 2022. pp. 133-146

[246] Joglekar S, Gajaralwar R. Potential risk and safety concerns of industrial nanomaterials in environmental management. In: Handbook of Nanomaterials for Wastewater Treatment. The Netherlands: Elsevier; 2021. pp. 1057-1079

[247] Cook L, Wantenaar C, Wise B. An exploratory study of the cutting-edge development of nanotechnology pertaining to the construction industry. Materials Science Forum. 2021;**1046**:95-103

[248] Zhu W, Bartos M. Application of nanotechnology in construction summary of a state-of-the-art report. 2004;**37**(November):649-658

Chapter 3

Advancing Sustainable Construction: Insights into Clay-Based Additive Manufacturing for Architecture, Engineering, and Construction

Maria De Los Angeles Ortega Del Rosario, Melany Medina, Rafael Duque, Antonio Alberto Jaén Ortega and Luis Castillero

Abstract

Additive manufacturing(AM) has evolved from rapid prototyping to a versatile technology in nano to large-scale fabrication, gaining traction in various sectors such as medicine, aeronautics, and pharmaceuticals. Its recent application in the architecture, engineering, and construction (AEC) industry marks a significant shift, especially in using traditional concrete and steel materials to innovative, sustainable options like clay. The increasing demand for ecofriendly construction materials propels this development. This chapter presents an overview of the latest developments in clay-based AM within the AEC sector. It discusses the challenges and opportunities of this technology, integrating design methods and material approaches. The chapter also examines the critical interplay of geometry, material properties, and process parameters in AM. Aimed at professionals in the field, it offers a comprehensive framework and practical guidelines for engineers, designers, and architects in this dynamic domain.

Keywords: clay, additive manufacturing, design for additive manufacturing, clay additive manufacturing, sustainability, architecture, building

1. Introduction

The architecture, engineering, and construction (AEC) sector is a complex and large industry that contributes to nearly 5–7% of the global gross domestic product (GDP) [1, 2], playing a significant role in the global workforce, economic growth and development [3]. Despite the economic significance and impact of the construction industry, it accounts for a substantial portion of global energy and material consumption and emissions. Specifically, the AEC industry accounts for 30% of the final energy demand, covering space heating and cooling, water heating, lighting, cooking, and other uses. This percentage could increase to 34% when the energy used to produce construction materials is also considered. Moreover, it represents 37% of energy-related CO_2 emissions and 21% of total greenhouse gas emissions [4]. Growing global concerns over climate change, the depletion of raw materials, and the substantial carbon and water footprints associated with conventional materials have catalyzed a vigorous and ongoing quest for strategies and technologies to mitigate these challenges and enhance the AEC industry's sustainability [5]. Consequently, there is a pronounced drive toward the adoption of innovative practices that not only reduce the environmental impact of construction activities but contribute to the resilience and sustainability of built environments. This encompasses exploring alternative materials, refining construction methodologies, and integrating eco-friendly technologies and designs to foster a more sustainable and environmentally responsible AEC industry.

Digital manufacturing and, furthermore, additive manufacturing (AM), commonly known as 3D printing in certain contexts, has been rapidly adopted by the AEC industry [6]. This technology spans a range of methodologies, each designed to enable the rapid, cost-effective, and efficient reproduction of components. The standards governing AM are encapsulated in ISO/ASTM 52900, which provides a comprehensive outline of the fundamental principles and terminologies associated with this domain. Thus, the AEC industry has widely adopted concrete-based additive manufacturing due to the attractive properties of the material, such as its mechanical performance, durability, availability, affordability, and workability, resulting in economic, technical, and scientific advantages [7–9], exploring concrete [10–13], cementitious and geopolymers [14–20]. Despite these advantages, the environmental footprint of using such materials remains a topic of concern [5, 9]. Their production and use still entail substantial energy consumption and CO_2 emissions, underscoring the need for continued research and development to pursue more sustainable construction methodologies and materials that align with environmental conservation goals. Moreover, using low-carbon and energy-efficient materials is paramount in decarbonization strategies [4], and thus, currently, it still exhibits great challenges to be compatible with the sustainability and resilience necessary in cities [21].

Using low-carbon and energy-efficient materials is paramount in decarbonization strategies [4], which remains a great challenge for concrete. Even if it is used coupled with AM, it is believed that it can increase the sustainability of the AEC industry since structures could be shaped in forms that can efficiently weigh between mechanical performance and material use, increasing productivity. Most cases in which sustainability is addressed and increased for concrete relate to the use of additives such as recycled and waste-sourced materials [22–24], geopolymers [20, 25], clay, and similar [9], that could reduce the emissions, the energy and the material use [26]. This scenario has spurred heightened interest in exploring alternative materials to concrete for construction through AM.

Therefore, clay has recently emerged as a material of significant interest in the context of AM within the construction industry. Despite the complexity of categorizing clays due to their diverse compositions and properties, they are broadly recognized as natural, fine-grained mineral materials. Clays exhibit a distinct plastic behavior when moist, allowing them to be easily shaped and molded. Upon drying or firing, clay hardens, transforming into a durable material suitable for various construction applications [27, 28]. This transformation, coupled with their availability and expected lower environmental impact, positions clay as a promising candidate for sustainable construction practices, especially when coupled with AM. The Literature currently reveals promising results in using clay with AM to create clay-based complex structures [29–35], which can be coupled, for instance, with natural fibers to increase their mechanical performance [36], promoting, for instance, responsive structures [37].

This chapter explores the use of clay as a construction material within the additive manufacturing AM context, an emerging focus in the AEC industry. It begins with a thorough review of significant advancements and case studies demonstrating clay's application in construction via AM technologies, underscoring its innovative potential in modern building practices. The chapter then delves into the crucial findings on how process parameters, material, and geometric parameters affect the performance and sustainability of clay-based construction. Drawing from recent research, it assesses the impact on structural integrity, thermal properties, and environmental impact. Finally, the chapter offers guidance and consolidates lessons learned from various projects, providing a framework for designers, architects, and engineers looking to use clay in sustainable construction through AM. This synthesis aims to deepen the understanding of clay's potential and promote its informed use in future projects.

2. Recent advances in clay-based additive manufacturing for the AEC industry

Since ancient times, humanity has used earth-based materials for construction, aiming to develop shelters and structures. These materials rank among the oldest used in construction due to their widespread availability, cost-effectiveness, environmental adaptability, and the extensive knowledge surrounding their use. Earth-based construction remains prevalent today, underscoring its enduring relevance and practicality. Recently, there has been a growing interest in integrating earth-based materials with AM technologies across various scales as a construction technique for civil and building elements within the AEC industry.

2.1 Clay-based AM houses

Among the most notable initiatives, the Institute for Advanced Architecture of Catalonia (IAAC) leads in large-scale additive manufacturing with clay. Their first project, Pylos, focused on refining techniques and testing mechanical properties via robotics. Additionally, IAAC's Digital Adobe wall, Terraperforma, and recent projects like TOVA and VOLTA prioritize sustainability, using local materials to achieve zero waste and minimal carbon emissions, showcasing the potential for sustainable construction practices within the AEC industry [38].

Wasp S.r.l developed some other well-known large-scale projects. For instance, they developed the Eremo project [39], which used a large-format Delta printer to develop a circular refuge model constructed from local materials, aiming for near-zero impact, and Gaia [40], the first circular-shaped, earth-based house made through additive manufacturing, that took just a couple of weeks to construct and requires no heating or air conditioning. The building material consisted of 25% local soil (30% clay, 40% silt, and 30% sand), 40% straw chopped rice, 25% rice husk, and 10% hydraulic lime. More recently, they also developed projects such as TECLA [41] using multiple crane-type Wasp 3D printers and collaboration with Dior to construct two circular modules installed in Dubai [42].

Another project developed during the COVID-19 pandemic was Casa Covida, a cohabitation concept in the San Luis Valley of the Colorado desert. This project integrated additive manufacturing with indigenous and traditional materials, differentiating itself from previous initiatives that used robotic and crane technologies. It employed a three-axis SCARA system, augmented with continuous flow and a stator-driven mortar pump, ensuring precise and efficient delivery of adobe material to the printing nozzle. This project also used natural-based and locally sourced building practices.

2.2 Envelopes and facades

An early study developed by Shi et al. [43] explored the potential of clay for additive manufacturing by introducing an interlocking screen system to develop a building envelope. Using parametric design techniques via Grasshopper, the team built each module with a Potterbot SCARA, using a commercial clay mix of 50:50 talc to ball clay. This modular approach was subsequently applied in the Hive project at Waterloo University, which featured a privacy wall consisting of 175 interlocking clay bricks [44].

Modulo Continuo, developed by Jiun Gan et al. [45], consisted of customized evaporative cooling façade modules. These modules were fabricated using a robotic arm with low-fire terracotta clay as the material. The authors tackled the intricacies of curved geometries by managing the convex deformation and curvature of the clay printings during the drying phase. The modules were printed by experimenting with various infill patterns and extruder pressures. By customizing each module's visual porosity and surface area, the authors suggested the potential for a highly tunable wall or façade, indicating a novel approach to architectural design that integrates functional esthetics with environmental needs.

More recently, Sanga et al. [46] applied biomimicry in designing additively manufactured clay bricks, drawing inspiration from termite mounds. Termites use their saliva as a binder to build underground shelters, breaking down consumed cellulose into glucose, polysaccharides, and oligosaccharides. Mirroring this process, the authors used cassava flour as a substitute to emulate the termite's construction behavior. Their findings showed that a 1.5% concentration of cassava flour in the mix yielded a compressive strength of 4.28 MPa, surpassing that of traditional burned clay bricks.

Furthermore, Taher et al. [47] delved into the design and fabrication of multi-functional building components, focusing on an approach to integrate air distribution duct networks directly into façade walls. They proposed a detailed workflow that employed parametric design and design for additive manufacturing (DfAM) principles to create a clay-based prototype. The study encountered several challenges related to geometric parameters, including the infill density and the complexity of intersection nodes along the toolpath. Despite these challenges, Taher et al. underscored the significant potential of clay for large-scale additive manufacturing within the AEC industry, highlighting its utility in developing innovative, functional building solutions.

2.3 Clay bricks

One of the most extended uses of clay in the AEC industry is bricks due to their easy-to-obtain load-bearing capacity as a structure. Economically, they are particularly appealing due to the low cost and widespread availability of raw materials, combined with the simplicity of the firing process used in their manufacture [48].

Despite a decline in their use since the 1980s in favor of concrete bricks, two-thirds of the global population is estimated to reside in buildings composed of clay materials [29, 48, 49]. The advent of digital fabrication technologies, particularly AM, has sparked renewed interest in clay bricks. This interest stems from AM's potential to enhance sustainability within the AEC industry. Early experiments in integrating AM with clay bricks were showcased during the Dutch Design Week through the Building Bytes project. This initiative featured a desktop 3D printer modified to produce clay bricks to create diverse architectural structures [50], demonstrating the evolving intersection between traditional materials and modern manufacturing techniques.

The first formal investigation into clay bricks 3D printing was conducted by Cruz et al., laying the groundwork for subsequent innovations in this field. Among these advancements, Abdallah and Estévez [29] introduced an innovative approach centered on physiological optimization for material deposition. This method is tailored to the properties of clay and employs a biodigital design methodology. Using Grasshopper, the authors devised a reaction-diffusion system that leverages the hydrophilic characteristics of clay. This approach aims to boost sustainability by reducing the material consumption and the energy required during brick fabrication through the development of an optimization algorithm that identifies the shortest path between two points during the construction process, ensuring a more efficient distribution of loads across the brick structure, thereby enhancing its resistance to cracking.

Building upon these foundations, Sangiorgio et al. [34] delved into parametric modeling, exploiting minimal surface and periodic minimal surface geometries with Grasshopper. This approach aims to design, simulate, and prototype complex bricks that leverage the exceptional mechanical properties of minimal surfaces. Their findings indicated that bricks structured around a diamond-type minimal surface yielded the most promising results. However, configurations such as the Scherk tower, gyroid, and Schwarz P also demonstrated favorable printability, illustrating the potential of parametric modeling in pushing the boundaries of clay brick manufacturing toward more intricate and mechanically robust designs.

2.4 Material approaches

While some efforts have successfully created complex structures, most have focused on developing the manufacturing technique and material performance. For instance, some research has focused on process parameters. An early investigation by Kontovourkis and Tryfonos aimed at developing an algorithm for robotic toolpath planning, focusing on parameters such as infill, overhang control, toolpath planning, robotic and nozzle control, and printing time using parametric design tools Rhino and Grasshopper [51, 52]. Pitayachaval and Baothong [53] developed a clay printing machine using a screw-based extrusion method to layer clay models through a circular nozzle. They focused on key process parameters such as nozzle diameter, screw extruder velocity, and screw pitch and introduced a mathematical model to correlate these variables, assessing their impact on the extrusion process. Their analysis showed that screw extruder velocity had a minimal effect on the performance across different nozzle diameters, providing insights into the interactions of these variables during the clay printing process.

Regarding the material key parameters, Bajpayee et al. [54] explored them using naturally harvested burlewash clay with a pH of 4.94, aiming to develop a load-bearing structure by crosslinking the clay through forming a siloxane framework to assess the feasibility of local sourcing. They crafted a mixture with a 1:1 weight ratio of sodium silicate and ground burlewash clay, enhanced with alkaline water and a cellulosic admixture to improve extrudability. Using Grasshopper and RobotStudio for robotic toolpath planning, they achieved a structure with a yield stress of 34 Pa and a kinematic viscosity of 26.4 Pa s at a shear rate of 1.4 s^{-1}, demonstrating the sustainable potential of this material in construction practices.

Sauter et al. [55] developed a cost-effective, mobile 3D-printing platform with omnidirectional wheels for unrestricted movement along x- and y-axes, designed primarily for fine arts rather than the AEC industry. Their study provided a statistical experimental analysis of process and material parameters, such as layer height, width, printing speed, and material composition, which included the ratio of dry material to wetting agent. They noted that the platform's mobility could introduce printing errors, particularly with larger objects, due to the lack of reference location. Similarly, Dielmans et al. [56] created a mobile robotic system for in-situ additive manufacturing (AM), testing its ability to produce accurate clay formworks for constructing reinforced, lightweight concrete columns.

Further research has focused on enhancing clay properties for AM. Sahoo and Gupta [57] integrated earth materials with alkali-activated slag-fly ash concrete, examining the rheological properties, extrudability, and buildability of mixtures containing 48–50% clay, predominantly Kaolinite with traces of Montmorillonite—with alkali-activated slag-fly ash concrete. While higher clay content presented structural strength challenges, their mixtures still met the IS1725 standards for soil-based block construction.

Research efforts have also been directed toward developing natural-based fibers to enhance clay-based mixtures for construction applications. Akemah and Ben-Alon [58] to identify suitable natural fibers for extrusion, aiming to optimize the mixtures for high static yield stress and minimal buckling, informed by research from Tay et al. [59]. They hypothesized that the effectiveness of a clay-based mixture as a construction material hinges on the cohesive interaction between the clay binder and the natural fiber reinforcement. Their experiments showed that adding fibers like wheat straw and hemp significantly improved the properties of clay. Evaluations confirmed that these fiber-enhanced clay mixtures are compatible with advanced additive manufacturing technologies such as direct and delta screw extrusion, demonstrating the potential of natural fibers to improve clay-based construction materials.

In another case, Jacquet and Perrot [60] explored the potential of enhancing clay by adding vegetable waxes, specifically coconut, soy, and rapeseed, focusing on soy wax. They discovered that a small addition of wax notably improved the water resistance of the clay materials, increasing their versatility. Furthermore, the wax's thermal properties facilitated a smoother manufacturing process by reducing the clay's stickiness and introducing a thermal setting aspect during printing. The study highlighted that these enhancements could benefit the AEC industry by providing innovative solutions for housing construction and reducing urban heat island effects. It also pointed out that the chemical composition of the materials influences the effectiveness of natural admixtures like wax and that the typical sticky and viscous nature of clay can be mitigated by adjusting the mixture composition or printing temperature.

3. Role of process, geometry, and material parameters

Earthenware and stoneware are the most common clay types used for architectural applications. The former consists of large particles and low-fire clay bodies, while the latter fires at high temperatures [61, 62]. Clay may differ from other materials commonly used for extrusion-based additive manufacturing in its unstable nature, mostly around its rheology and other properties, and the influence of the pumping pressure on the quality of the printed objects [63].

Even though the use of clay in additive manufacturing has been increasing recently, there remains a notable deficiency in systematic information, standardization, and testing to evaluate process, material, and geometry parameters. Directly applying standards common to traditional clay use could be challenging and may result in inaccurate assessments [5, 64]. Although the research addressing these parameters is not extensive, several key effects have been documented across various studies. This section provides a reference framework, summarizing significant findings and establishing a foundation for future research. This summary aims to consolidate knowledge and identify critical areas where further investigation could enhance understanding and improve practices in the relevant disciplines.

3.1 Permeability

Permeability is a crucial parameter often used to characterize soils, including clays. The permeability of clay is influenced by several factors, such as its particle structure, shape, pore size, uniformity and distribution, high specific surface area, the strong attraction of water to clay particle surfaces, and its plasticity [65–67]. Due to its cohesive nature, clay is generally considered to have low permeability. Various empirical and theoretical relationships, such as the Atterberg limits, have been established to determine clay permeability. These limits are used for the identification, description, and classification of clays and to evaluate their mechanical properties [65, 68]. One can also mention the extensively used Kozeny-Carman equation, an empirical approach widely used to determine the specific surface area of cohesive soils by predicting the hydraulic conductivity of soils [69].

The permeability of clays contributes to their hygric capacity, which can be leveraged to regulate indoor temperature and humidity levels, offering potential benefits in the AEC industry [58]. In the case of concrete, permeability in 3DCP can be assessed through typical experiments such as mercury intrusion porosimetry (MIP), water absorption, and sorptivity. However, the Literature has not extensively explored the effects of permeability on clay during and after the 3D printing process.

A recent study by Carr et al. [66] investigated the impact of 3D printing on the physical characteristics of clay, covering aspects like Atterberg limits, particle size gradation, and specific gravity. This study is among the few that have explored the effects of permeability in clay 3D printing, establishing a valuable framework for future research in this area. In their experiments, cylindrical clay-based specimens with a diameter of 35.5 mm and height of 50.8 mm, weighing 50 g each, were fabricated using Kaolinite clay with a void ratio of $e = 1.0$. To enhance the flowability of the powder, it was dried for 10 hours at 200°C to minimize water content during the AM process. After printing, these specimens were sintered at 1093°C. For comparative analysis, control specimens were prepared using the dry pluviation technique with stoneware clay powder, achieving a void ratio of e=1.0, then remolded to facilitate direct comparison with the 3D printed specimens.

The authors addressed the impact of 3D printing on soil properties, focusing on clay permeability and shear strength. Sieve analysis was conducted on pulverized 3D-printed specimens to assess their particle size distribution, identifying the clay as low-plasticity. Analysis revealed particle fusion during sintering, and despite efforts to meet Atterberg limits through 24-hour hydration with de-aired water, these were not achieved. This suggested that the stoneware clay powder did not maintain cohesiveness post-printing. Further insights were obtained using scanning electron microscopy (SEM), which showed significant changes in the particle structures post-printing, notably smoother surface textures. This textural change, likely due to the addition of fire-resistant materials, was hypothesized to enhance specimen permeability by increasing porosity and reducing shrinkage during sintering, potentially improving the clay's permeability. Nonetheless, permeability variability was minor and less influenced by confining pressure. The analysis also indicated increased strength in the printed specimens.

3.2 Plasticity and extrudability

The plasticity of clay-based materials is significantly influenced by their water content, a factor critical to their extrudability, especially in 3D printing applications [70]. Kaolin clay, for instance, features a crystalline platelet-like structure comprised of small particles and water of formation. In this context, water functions similarly to a lubricant, allowing particles to glide over each other under shear forces rather than breaking apart. This mechanism enables kaolin clay to exhibit plastic shear-thinning behavior, known as thixotropy, where the surface tension between water and clay particles helps maintain the object's shape. However, water content also impacts the flowability of the clay. If the water content surpasses the minimum needed to sustain surface tension, the clay becomes overly fluid, consequently losing its ability to hold a defined shape [71]. Balancing water content and plasticity is crucial for successful 3D printing with clay, where maintaining the right consistency is critical to achieving the desired structural integrity and shape.

3.3 Buildability

Buildability is one of the most common criteria for testing the performance of an additively manufactured part, especially for cement-based and clay-based materials [57, 58, 64, 72]. This property can be evaluated both qualitatively and quantitatively, focusing on factors such as the build rate and the stability of the printed layers. Critical aspects that influence buildability include the maximum load that the initial layer can support, the material's early strength development, and the printed part's geometry. These factors are crucial in determining how effectively the structure can withstand the successive layering process without undergoing deformation or collapse, thereby ensuring that the additive manufacturing process yields durable and structurally sound components. Despite its importance, standardization in buildability testing remains underdeveloped. However, several tests are referenced in the Literature, such as the cylinder and single-direction wall tests, commonly used to assess the structural integrity and layer cohesion of 3D-printed objects. Additionally, stability settlement tests are also mentioned, which evaluate the plastic collapse of a structure under the maximum load that the first layer can bear.

3.4 Flow rate

Flow rate measurements are critical in optimizing the extrusion process in additive manufacturing. For instance, Fleck et al. [33] examined how the flow rate of material exiting the nozzles correlates with the temperature at the nozzle tips. In their study, temperature increases were induced by vibration, which, in turn, exponentially increased the flow rate. The authors employed the Andrade equation to model this behavior, which connects viscosity with temperature under the assumption of Bingham plastic behavior. This equation was paired with the relationship of laminar flow rate to viscosity, suggesting that the increased flow rates resulted from reduced viscosity due to the rise in temperature and vibrations at the nozzle tip.

The flow rate of extrusion also significantly influences other printing parameters. Wi et al. [35] noted that if the extrusion flow rate is too slow or too fast relative to the set printing speed, it can adversely affect the printability and buildability of the object. Inappropriate flow rates may cause the object to tilt, slump, or collapse over time, leading to severe distortion or poor printing quality.

Moreover, some studies have approached numerically the prediction of flow rates during additive manufacturing of cementitious and ceramic materials, but there is still a lack of studies on clay additive manufacturing.

3.5 Surface roughness

The surface roughness of 3D printed clay parts is a critical aspect often assessed by non-destructive methods such as direct visual inspection through images [53, 73] and a light scanning system [35]. These methods provide valuable insights into the effects of printing parameters, geometry, and material characteristics on the final product. Surface roughness is a significant parameter in manufacturing industries because it is closely related to product quality and precision, impacting critical factors such as friction, contact deformation, heat and electric current conduction, the tightness of contact joints, and positional accuracy [74]. Pitayachaval and Baothong [53] noted that higher screw pitches tend to result in smoother surfaces, although their assessments were purely qualitative without further quantitative measurements. Conversely, Wi et al. [35] introduced a 3D Light Scanning System (3D-SLSS) that operates akin to human stereo vision but uses a projector to emit multiple phase-shift patterns sequentially while a camera captures the surface images of the object. This setup allows for non-contact, non-destructive testing of various visual and surface parameters, including surface roughness. For this study, surface roughness was quantified using metrics such as the arithmetic average height (Ra), root mean square roughness (Rq), and 10-point height (Rz), finding that roughness increased with the extrusion flow rate and printing speed.

Additionally, Afriat et al. [73] explored the quality of external surfaces on printed elements using vibration-assisted printing (VAP), which allows the use of smaller nozzle diameters. This approach resulted in improved surface quality and sharper corner angles. The surface quality was evaluated using GOM Inspect (now ZEISS INSPECT Optical 3D), enhancing the assessment of the printed elements' external characteristics. These methodologies highlight the importance of precise control over printing parameters to optimize surface quality in 3D-printed clay products.

3.6 Layer thickness

Layer thickness is a critical parameter directly impacting the printed part's surface roughness and overall quality. Smaller extruder diameters are advantageous as they tend to produce thinner layers, resulting in smoother surface finishes. Some results in the Literature have shown an interesting relationship where the higher the printing speed, the less viscous the fluid becomes [71]. This observation aligns with the behavior of supersaturated clay suspensions, which are known to exhibit shear-thinning properties. This thixotropic, non-Newtonian behavior of clay makes it particularly suitable for processes where control over viscosity and flow rate is essential for achieving high-quality prints.

Besides, the first layer is fundamental for the printing process. Among the various factors influencing the first layer in printing, nozzle parameters, printing speed, bed temperature, surface adhesion, bed leveling, and the Z-offset calibration of the printer are crucial [33, 71]. Boyer et al. [71] investigated the impact of three different substrate types on the printing process: aluminum, paper, and polishing paper. They found that the material tended to slide over the surface of aluminum, which could compromise print quality. Although initial adhesion was adequate with paper substrates, the quality deteriorated over time, likely due to the paper's water absorbance. In contrast, polishing paper, characterized by lower permeability and higher surface roughness, provided better adhesion and overall results than the other substrates tested. This study highlights the importance of selecting the appropriate substrate to ensure optimal first-layer quality and overall printing success.

3.7 Nozzle geometry

The nozzle is a critical component of material extrusion-based additive manufacturing technologies, serving as the conduit through which the material is extruded. Its design and geometry must be meticulously tailored to align with the specific properties of the material used, including its composition and consistency. Key factors such as viscosity, the presence of aggregates, and fiber lengths must be considered to ensure that the nozzle can effectively and efficiently handle the material without clogging or causing irregular extrusion. For cement-based and clay-based materials, the nozzle shape usually follows two types of sections: rounded or elliptical and rectangular or squared [75, 76], which are associated with the printing quality [5], including factors such as the presence of air pockets, cold joints, adherence between layers, among others [77].

The nozzle diameter is a key parameter in additive manufacturing, requiring careful selection to meet specific requirements. According to Chan et al. [32], the choice of nozzle significantly affects the printed parts' quality, speed, material consumption, and precision. It is feasible to use larger nozzles provided they are paired with suitable combinations of printing pressures and speeds once the minimum nozzle size is exceeded. Furthermore, the design and architecture of the nozzle itself are crucial for the quality of the manufactured part.

Jauk et al. [78] introduced customized nozzles for the Delta WASP 40100 Clay model printer, designed to enhance the printing of filament-reinforced clay objects. These nozzles feature a filament-guiding mechanism that centers the filament within the extrusion channel, ensuring consistent application. The nozzle design includes a threaded base for secure attachment to the extruder, a smaller inner nozzle to guide the filament, and a larger outer nozzle that shapes the extruded clay. This adaptation is specifically tailored to incorporate fiber reinforcement directly into the clay matrix during printing, potentially increasing the tensile strength of the final product by approximately 15%, irrespective of the fiber type used. This innovation exemplifies how adjustments to nozzle design can significantly enhance the functional properties of 3D-printed objects.

The printer nozzle plays a crucial role in determining the resolution and detail of printed parts, with smaller nozzles allowing for thinner layers and greater accuracy in rendering fine details. According to the results shown by Chan et al. [32], a decrease in pressure and velocity is observed as the nozzle size increases. Conversely, increasing the storage space while maintaining the same printing speed and nozzle size needs higher pressure for effective printing. If the nozzle diameter is large, it results in higher material consumption per layer, which can significantly impact production costs. Conversely, smaller nozzle sizes enhance material efficiency, minimizing waste. Fleck et al. [33] delve into the critical parameters of extrusion additive manufacturing that are essential for maintaining control over material geometry, especially when printing at high viscosities and variable speeds, which can lead to over-extrusion.

Additionally, their rheology analysis showed that the printing was unsuccessful for diameters lower than approximately 0.26–0.30 mm. This suggests that the viscosity of the material is a limiting factor in achieving successful printing under certain conditions. Furthermore, larger nozzle diameters can facilitate higher printing speeds, allowing more material to be extruded quickly. This capability is particularly advantageous for the rapid production of parts. Afriat et al. [73] highlight an additional benefit when employing vibration-assisted printing (VAP), which can achieve printing speeds ranging from 5000 to 6000 mm/min. This significant increase in printing speed substantially reduces the time required to print highly viscous materials and enables the extrusion of even more viscous substances. The interplay between nozzle size, material viscosity, and printing parameters like pressure and speed are critical to optimizing 3D printing outcomes.

3.8 Tensile behavior

In the case of Fleck et al. [33], the authors assessed the mechanical properties of 3D-printed dogbones, adhering to ASTM D638 standards, and compared them with those manufactured conventionally. The findings indicated that the 3D-printed parts generally displayed poorer mechanical performance. Notably, parts printed at 0° orientation exhibited the highest elastic modulus, recorded at 6.10 ± 0.68, while the greatest ultimate strength was observed in parts printed at 90°, achieving 1.29 ± 0.07. The unexpected strength increase in 3D-printed parts at a 90° orientation was surprising, as this alignment is typically weaker due to its perpendicular orientation to the pulling direction, akin to composite laminates. However, the authors suggest that overfilling during printing may have inadvertently increased the interface's contact pressure, thus enhancing the ultimate strength. This finding suggests that the interaction between material deposition and mechanical stresses during printing significantly impacts the final properties of 3D-printed objects, indicating a need for further research to optimize printing strategies for improved mechanical outcomes. Additionally, it highlighted that corner-turning in print geometry could cause overfilling, leading to reduced geometric accuracy due to the firmware's requirement for zero-velocity changes based on Marlin's trapezoidal velocity profiles [79]. To mitigate this, adjustments in the g-code are necessary to manage the flow and accommodate temperature-induced material behavior.

3.9 Printing quality and costs

Print speed is a crucial factor that significantly impacts operating costs in additive manufacturing. Faster printing speeds can reduce energy and material consumption, offering economic benefits. According to Kim et al. [80], the screw-type extrusion method is versatile, capable of handling various material types, including very thick ones, and can complete numerous tasks quickly. However, this method faces challenges in achieving precision, particularly with complex shapes. Increasing the printing speed initially may reduce the production time but at the cost of the quality of the output. High printing speeds are often associated with imperfections in the final part, such as layer errors or reduced resolution. Sauter et al. [55] delve into this issue by investigating the optimal printing parameters that minimize errors. Their study provides a detailed analysis of various parameter pairings such as layer width and concentration, layer height and concentration, layer width and printing speed, layer height and printing speed, and printing speed and concentration.

4. Lessons, challenges, and opportunities

One significant barrier to the broader adoption of clay in AM is the lack of clear guidelines, largely due to limited research on essential parameters like geometry, material, and process. Although many researchers and companies have demonstrated its technical feasibility for various applications such as facades, these studies often emphasize broad viability over detailed operational specifics.

This chapter addresses the need for more targeted research to understand how these key parameters impact outcomes in AM processes using clay. The AEC industry can harness AM with clay more effectively to enhance sustainability by developing a deeper understanding of these interactions. This could lead to more robust guidelines and standards that facilitate the integration of clay-based AM into mainstream construction practices, improving efficiency, cost-effectiveness, and environmental impact. Different approaches are essential to maximize the potential of AM, particularly when tailored to specific applications. Like traditional manufacturing, AM performance is influenced by various factors, focusing on technical, economic, and sustainability aspects to determine the most suitable method for the intended application. One strategy for leveraging AM's full potential that considers the design of parts and products within the constraints of their capabilities is design for additive manufacturing (DfAM) [81].

DfAM is part of the broader DfX (design for X) family of methods, which aims to ensure that the engineering design process aligns effectively with AM capabilities. This approach helps to harness techno-economic benefits such as the ability to produce low-volume, high-complexity parts efficiently. DfAM involves using advanced tools that cater specifically to AM design challenges. DfAM must address the various technologies encompassed within AM and attempt to integrate them into a cohesive process chain. This integration allows all manufacturing process components to interact effectively, leading to efficient production outcomes [82]. By focusing on these aspects, DfAM optimizes the manufacturing process and enhances the functionality and customization of the products, making it a pivotal approach in the evolution of additive manufacturing.

The use of DfAM typically involves establishing a reference framework that facilitates the exploration, design, optimization, and manufacturing of parts or products [83–87]. For this chapter, the application of DfAM focused primarily on developing process parameters, geometry, and material approaches. In this context, the analysis deliberately excludes broader environmental, economic, and social factors, concentrating solely on the technical aspects that directly influence the various stages of design and manufacturing within AM. This focused approach allows for a deeper understanding and refinement of the technical factors critical to optimizing additive manufacturing processes. By narrowing the scope of investigation, this chapter aims to provide clear, actionable insights that enhance the effectiveness and efficiency of DfAM in addressing complex technical challenges within the AM field.

Based on a holistic analysis of data available in the literature and from manufacturers, the process parameters, geometry, and materials discussed in the previous sections were used to define a tailored framework for DfAM. This framework, inspired by the methodology outlined by Trovato and Cicconi [88], is designed to cater specifically to the unique challenges and opportunities presented in this particular case. The details of this adapted framework are visually represented and can be thoroughly examined in **Figure 1**. This figure illustrates how the conceptual framework integrates various elements of DfAM to effectively address the practical aspects of additive manufacturing, ensuring that the design process is optimized for the specific materials and technologies involved.

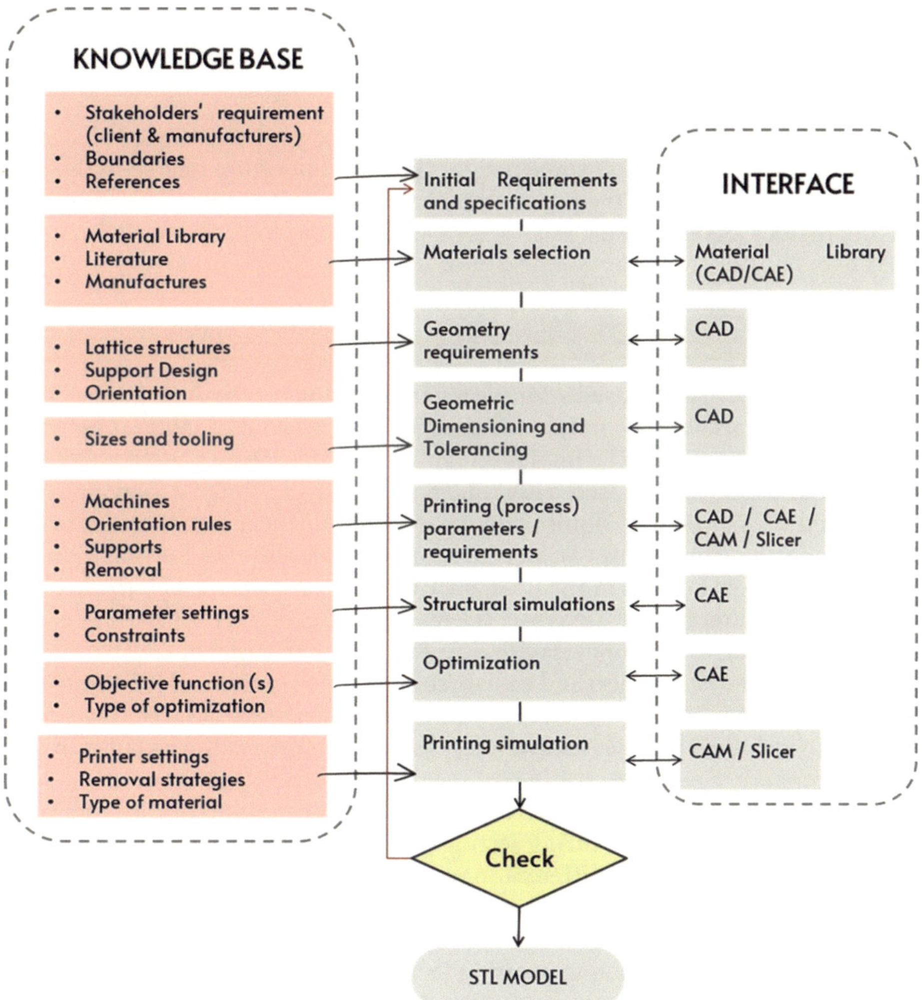

Figure 1.
Proposed DfAM reference framework for manufacturing clay-based non-structural blocks, adapted from the methodologies proposed by Trovato and Cicconi [88].

Issue	Possible cause and proposed solutions	References
Lines and marks in the direction of extrusion	Cause: issues with the nozzle (adhered particles, damage, and others) or inadequate processing conditions. Proposal: clean the nozzle. Check for any damage.	[73] [31] [32]
Non-uniform thickness in the transverse direction	Cause: nozzle calibration Proposal: calibrate the nozzle to ensure it is centered.	[33] [89] [80]
Non-uniform thickness in the longitudinal direction	Cause: flow fluctuations. Proposal: check for the absence of pressure or motor power oscillations. Verify the extruder temperature. Ensure the temperature controllers, resistors, and fans are functioning properly.	[32] [33] [89] [80]
Presence of impurities	Cause: adhesion in the nozzle, head, filters, or even extruder. Inadequate processing conditions. Proposal: check the condition of filters. Clean the head, nozzle, or extruder. Purge with another material. Check for potential sources of contamination. Validate and adjust processing conditions if necessary.	[31] [53]

Table 1.
Common problems and explored solutions reported in the literature.

Furthermore, **Table 1** shows some insights into the Literature related to the process, material, and geometry parameters that can serve as a troubleshooting table based on previous experiences.

5. Conclusions

Recent advances in clay-based additive manufacturing for the AEC industry have demonstrated remarkable progress, reflecting a growing trend of integrating traditional materials with modern technological approaches to enhance sustainability and efficiency. The use of earth-based materials, notably clay, in construction dates back millennia and continues to be significant due to its abundance, cost-effectiveness, and versatile properties suitable for various environmental conditions. Several pioneering projects have set benchmarks in the field, such as those spearheaded by the Institute for Advanced Architecture of Catalonia (IAAC). These projects have used large-scale additive manufacturing techniques with clay, enhancing traditional construction methods with modern precision and sustainability considerations. However, the application of this technology extends beyond the construction of large structures like houses to include envelopes and facades, as well as more traditional uses in brick manufacturing.

Research in this field is rapidly increasing, with studies exploring different properties of clay in additive manufacturing settings that can provide a deeper understanding of how clay behaves under different manufacturing conditions and how these properties can be optimized to improve product quality and structural integrity. The field of clay-based additive manufacturing is poised to expand further, exploring new material combinations, improving printing technologies' precision, and enhancing construction practices' sustainability. Collaborative efforts between academic institutions, industry leaders, and architectural firms are crucial in driving innovation and adoption of these technologies in mainstream construction projects. By bridging the gap between traditional materials and advanced manufacturing techniques, the AEC industry can better address the growing demands for sustainable development and environmentally adaptive construction solutions.

Acknowledgements

The authors sincerely thank the Design, Materials, and Manufacturing Research Group (DM+M) at the Faculty of Mechanical Engineering, Universidad Tecnológica de Panamá (UTP), for their invaluable collaboration. This work was partially supported by projects FIED22-13 and APY-NI-2022-24, both funded by the Panamanian Secretaría Nacional de Ciencia, Tecnología e Innovación (SENACYT), and by the Sistema Nacional de Investigación (SNI).

Conflict of interest

The authors declare no conflict of interest.

Author details

Maria De Los Angeles Ortega Del Rosario[1,2,3]*, Melany Medina[1], Rafael Duque[1], Antonio Alberto Jaén Ortega[1,4] and Luis Castillero[1]

1 Faculty of Mechanical Engineering, Research Group in Design, Manufacturing and Materials (DM+M), Universidad Tecnológica de Panamá, Panama City, Panama

2 Centro de Estudios Multidisciplinarios en Ciencias, Ingeniería y Tecnología (CEMCIT-AIP), Panama City, Panama

3 Sistema Nacional de Investigación (SNI), Panama City, Panama

4 Polymer Chemistry and Biomaterials Research Group, Centre of Macromolecular Chemistry (CMaC), Ghent University, Belgium

*Address all correspondence to: maria.ortega@utp.ac.pa

References

[1] Adepu N, Kermanshachi S, Pamidimukkala A, Loganathan K. Analyzing the factors affecting construction project schedules amidst COVID-19 pandemic. Project Leadership and Society. 2023;**4**:100100

[2] Pan Y, Zhang L. Roles of artificial intelligence in construction engineering and management: A critical review and future trends. Automation in Construction. 2021;**122**:103517

[3] Arowoiya VA, Moehler RC, Fang Y. Digital twin technology for thermal comfort and energy efficiency in buildings: A state-of-the-art and future directions. Energy and Built Environment. 1 Oct 2024;**5**(5):641-656. DOI: 10.1016/s0014-5793(01)03293-8

[4] United Nations Environment Programme. Global Status Report for Buildings and Construction: Beyond foundations: Mainstreaming sustainable solutions to cut emissions from the buildings sector. Nairobi. United Nations Environment Programme (UNEP) and Global Alliance for Buildings and Construction (GlobalABC). 2024. DOI: 10.59117/20.500.11822/45095

[5] Rocha D, Faria P, Lucas SS. Additive manufacturing of earth-based materials: A literature review on mortar composition, extrusion, and processing earth. Materials. 2024;**17**(1):202

[6] Teixeira J, Schaefer CO, Rangel B, Maia L, Alves JL. A road map to find in 3D printing a new design plasticity for construction—The state of art. Frontiers of Architectural Research. 2023;**12**(2):337-360

[7] Dilawar Riaz R, Usman M, Ali A, Majid U, Faizan M, Jalil MU. Inclusive characterization of 3D printed concrete (3DPC) in additive manufacturing: A detailed review. Construction and Building Materials. 2023;**394**:132229

[8] Rakowska J, Węgrzyn M, Rudnik E. Impact of ionic liquids on absorption behaviour of natural fibers/biopolyethylene biocomposites. Scientific Reports. 2021;**11**(1):20483

[9] Motalebi A, Khondoker MAH, Kabir G. A systematic review of life cycle assessments of 3D concrete printing. Sustainable Operations and Computers. 2024;**5**:41-50

[10] Ahmed GH. A review of "3D concrete printing": Materials and process characterization, economic considerations and environmental sustainability. Journal of Building Engineering. 1 May 2023;**66**:105863

[11] Mohan MK, Rahul AV, De Schutter G, Van Tittelboom K. Extrusion-based concrete 3D printing from a material perspective: A state-of-the-art review. Cement and Concrete Composites. 2021;**115**:103855

[12] Nguyen-Van V, Nguyen-Xuan H, Panda B, Tran P. 3D concrete printing modelling of thin-walled structures. Structure. 2022;**39**:496-511

[13] Wu Y, Liu C, Liu H, Zhang Z, He C, Liu S, et al. Study on the rheology and buildability of 3D printed concrete with recycled coarse aggregates. Journal of Building Engineering. 2021;**42**:103030

[14] Adamu M, Alanazi F, Ibrahim YE, Alanazi H, Khed VC. A comprehensive review on sustainable natural fiber in cementitious composites: The date palm fiber case. Sustainability. 2022;**14**(11):6691

[15] Aydin EM, Kara B, Bundur ZB, Ozyurt N, Bebek O, Ali GM. A comparative evaluation of sepiolite and nano-montmorillonite on the rheology of cementitious materials for 3D printing. Construction and Building Materials. 2022;**350**:128935

[16] Ilcan H, Sahin O, Kul A, Yildirim G, Sahmaran M. Rheological properties and compressive strength of construction and demolition waste-based geopolymer mortars for 3D-printing. Construction and Building Materials. 2022;**328**:127114

[17] Korniejenko K, Frączek E, Pytlak E, Adamski M. Mechanical properties of geopolymer composites reinforced with natural fibers. Procedia Engineering. 2016;**151**:388-393

[18] Nodehi M, Ozbakkaloglu T, Gholampour A. Effect of supplementary cementitious materials on properties of 3D printed conventional and alkali-activated concrete: A review. Automation in Construction. 2022;**138**:104215

[19] Al-Ghazali NA, Aziz FNAA, Abdan K, Nasir NAM. Mechanical Properties of Natural Fibre Reinforced Geopolymer Composites: A Review. Pertanika Journal of Science & Technology. Jul 2022;**30**(3):2053-2069. DOI: 10.47836/pjst.30.3.16

[20] Raza MH, Zhong RY. A sustainable roadmap for additive manufacturing using geopolymers in construction industry. Resources, Conservation and Recycling. 2022;**186**:106592

[21] Dong L, Wang Y, Li HX, Jiang B, Al-Hussein M. Carbon reduction measures-based LCA of prefabricated temporary housing with renewable energy systems. Sustainability. 2018;**10**(3):718

[22] Adaloudis M, Bonnin RJ. Sustainability tradeoffs in the adoption of 3D concrete printing in the construction industry. Journal of Cleaner Production. 2021;**307**:127201

[23] Liu S, Lu B, Li H, Pan Z, Jiang J, Qian S. A comparative study on environmental performance of 3D printing and conventional casting of concrete products with industrial wastes. Chemosphere. 2022;**298**:134310

[24] Zhou Y, Luo H, Anand K, Singh A, Xie YM. Sustainable use of ultrafine recycled glass in additive manufactured (3D printed) reactive powder concrete. Construction and Building Materials. 2024;**419**:135556

[25] Demiral NC, Ozkan Ekinci M, Sahin O, Ilcan H, Kul A, Yildirim G, et al. Mechanical anisotropy evaluation and bonding properties of 3D-printable construction and demolition waste-based geopolymer mortars. Cement and Concrete Composites. 2022;**134**:104814

[26] Şahin O, İlcan H, Ateşli AT, Kul A, Yıldırım G, Şahmaran M. Construction and demolition waste-based geopolymers suited for use in 3-dimensional additive manufacturing. Cement and Concrete Composites. 2021;**121**:104088

[27] Bergaya F, Lagaly G. Chapter 1—General introduction: Clays, clay minerals, and clay science. In: Bergaya F, Lagaly G, editors. Developments in Clay Science, Handbook of Clay Science. Vol. 5. Elsevier; 2013. pp. 1-19. Available from: https://www.sciencedirect.com/science/article/pii/B9780080982588000018

[28] Guggenheim S, Martin RT. Definition of clay and clay mineral: Joint report of the Aipea nomenclature and CMS nomenclature committees. Clays and Clay Minerals. 1995;**43**(2):255-256

[29] Abdallah YK, Estévez AT. 3D-printed biodigital clay bricks. Biomimetics. 2021;**6**(4):59

[30] Alonso Madrid J, Sotorrío Ortega G, Gorostiza Carabaño J, Olsson NOE, Tenorio Ríos JA. 3D claying: 3D printing and recycling clay. Crystals. 2023;**13**(3):375

[31] Chaari MZ, Abdelfatah M, Loreno C. A trial to convert a polymer FDM 3D printer to handle clay materials. SN Applied Sciences. 2022;**4**(3):68

[32] Chan SSL, Pennings RM, Edwards L, Franks GV. 3D printing of clay for decorative architectural applications: Effect of solids volume fraction on rheology and printability. Additive Manufacturing. 2020;**35**:101335

[33] Fleck TJ, McCaw JCS, Son SF, Gunduz IE, Rhoads JF. Characterizing the vibration-assisted printing of high viscosity clay material. Additive Manufacturing. 2021;**47**:102256

[34] Sangiorgio V, Parisi F, Fieni F, Parisi N. The new boundaries of 3D-printed clay bricks design: Printability of complex internal geometries. Sustainability. 2022;**14**(2):598

[35] Wi K, Suresh V, Wang K, Li B, Qin H. Quantifying quality of 3D printed clay objects using a 3D structured light scanning system. Additive Manufacturing. 2020;**32**:100987

[36] Robayo-Salazar R, Mejía De Gutiérrez R, Villaquirán-Caicedo MA, Delvasto AS. 3D printing with cementitious materials: Challenges and opportunities for the construction sector. Automation in Construction. 2023;**146**:104693

[37] Ortega Del Rosario MDLÁ, Beermann K, Chen AM. Environmentally responsive materials for building envelopes: A review on manufacturing and biomimicry-based approaches. Biomimetics. 2023;**8**(1):52

[38] Institute for Advanced Architecture of Catalonia (IAAC). Projects & Development Institute for Advanced Architecture of Catalonia (IAAC). IAAC; 2024. Available from: https://iaac.net/projects-development/

[39] Moretti F. 3D Printed Houses for a Renewed Balance Between Environment and Technology. 3D Printers | WASP; 2017. Available from: https://www.3dwasp.com/en/3d-printed-houses-for-a-renewed-balance-between-environment-and-technology/

[40] Chiusoli A. The First 3D Printed House with Earth | Gaia. 3D Printers | WASP; 2018. Available from: https://www.3dwasp.com/en/3d-printed-house-gaia/

[41] Chiusoli A. 3D Printed House TECLA—Eco-Housing. 3D Printers | WASP; 2021. Available from: https://www.3dwasp.com/en/3d-printed-house-tecla/

[42] Moretti F. A 3D Printed Pop-Up Store by WASP for Dior. 3D Printers | WASP; 2021. Available from: https://www.3dwasp.com/en/3d-printed-pop-up-store-wasp-dior/

[43] Shi J, Cho Y, Taylor M, Correa D. Guiding instability a craft-based approach for modular 3D clay printed masonry screen units. In: Blucher Design Proceedings. Porto, Portugal: Editora Blucher; 2019. pp. 477-484. Available from: http://www.proceedings.blucher.com.br/article-details/34204

[44] Caldwell B. Printing with wet clay: Architecture team produces showpiece privacy wall for high-rise office using 3D-printed bricks. Waterloo News. 2022. Available from: https://uwaterloo.ca/news/printing-wet-clay

[45] Gan AWJ, Guida G, Dongyum K, Devashree S, Hyejun Y, Zach S. Modulo

Continuo—5-axis Ceramic Additive Manufacturing Applications for Evaporative Cooling Facades Modules. In: Co-creating the Future: Inclusion in and through Design - Proceedings of the 40th Conference on Education and Research in Computer Aided Architectural Design in Europe (eCAADe 2022) [Internet]. Leuven, Belgium: CUMINCAD; 2022 [cited 2024 May 20]. p. 47-55. Available from: https://papers.cumincad.org/cgi-bin/works/paper/ecaade2022_275

[46] Sanga R, Kilumile M, Mohamed F. Alternative clay bricks inspired from termite mound biomimicry. Case Studies in Construction Materials. 2022;**16**:e00977

[47] Taher A, Aşut S, van der Spoel W. An integrated workflow for designing and fabricating multi-functional building components through additive manufacturing with clay. Buildings. 2023;**13**(11):2676

[48] Lozano-Miralles JA, Hermoso-Orzáez MJ, Martínez-García C, Rojas-Sola JI. Comparative study on the environmental impact of traditional clay bricks mixed with organic waste using life cycle analysis. Sustainability. 2018;**10**(8):2917

[49] Mary Lissy PN, Peter C, Mohan K, Greens S, George S. Energy efficient production of clay bricks using industrial waste. Heliyon. 2018;**4**(10):e00891

[50] Etherington R. Building bytes 3D printed bricks by Brian Peters at Dutch design week. Dezeen. 2012. Available from: https://www.dezeen.com/2012/10/31/building-bytes-3d-printed-bricks-brian-peters/

[51] Kontovourkis O, Tryfonos G. Integrating parametric design with robotic additive manufacturing for 3D clay printing: An experimental study. In: Proceedings of the 35th ISARC, Berlin, Germany: The International Association for Automation and Robotics in Construction; 2018. pp. 918-925. Available from: https://www.iaarc.org/publications/2018_proceedings_of_the_35th_isarc/integrating_parametric_design_with_robotic_additive_manufacturing_for_3d_clay_printing-an_experimental_study.html

[52] Kontovourkis O, Tryfonos G. Robotic 3D clay printing of prefabricated non-conventional wall components based on a parametric-integrated design. Automation in Construction. 2020;**110**:103005

[53] Pitayachaval P, Baothong T. An effect of screw extrusion parameters on a pottery model forming by a clay printing machine. International Journal of Integrated Engineering. 2022;**14**(6):38-46

[54] Bajpayee A, Farahbakhsh M, Zakira U, Pandey A, Ennab LA, Rybkowski Z, et al. In situ resource utilization and reconfiguration of soils into construction materials for the additive manufacturing of buildings. Frontiers in Materials. 2020;**7**:52. Available from: https://www.frontiersin.org/articles/10.3389/fmats.2020.00052

[55] Sauter A, Nasirov A, Fidan I, Allen M, Elliott A, Cossette M, et al. Development, implementation and optimization of a mobile 3D printing platform. Progress in Additive Manufacturing. 2021;**6**(2):231-241

[56] Dielemans G, Lachmayer L, Recker T, Atanasova L, Hechtl CM, Matthäus C, et al. Mobile additive manufacturing: A case study of clay formwork for bespoke in situ concrete construction. In: Buswell R, Blanco A, Cavalaro S, Kinnell P, editors. Third RILEM International Conference on

Concrete and Digital Fabrication. Cham: Springer International Publishing; 2022. pp. 15-21

[57] Sahoo P, Gupta S. 3D printable earth-based alkali-activated materials: Role of mix design and clay-rich soil. In: Amziane S, Merta I, Page J, editors. Bio-Based Building Materials. Cham: Springer Nature Switzerland; 2023. pp. 333-352

[58] Akemah T, Ben-Alon L. Developing 3D-printed natural fiber-based mixtures. In: Amziane S, Merta I, Page J, editors. Bio-Based Building Materials, RILEM Book Series. Cham: Springer Nature Switzerland; 2023. pp. 555-572

[59] Tay YWD, Qian Y, Tan MJ. Printability region for 3D concrete printing using slump and slump flow test. Composites Part B: Engineering. 2019;**174**:106968

[60] Jacquet Y, Perrot A. Evolutionary Approach Based on Thermoplastic Bio-Based Building Material for 3D Printing Applications: An Insight into a Mix of Clay and Wax. In: Amziane S, Merta I, Page J, editors. Bio-Based Building Materials Proceedings of ICBBM 2023. Switzerland: Springer Nature Switzerland; 2023. pp. 271-279

[61] Martín-Márquez J, Rincón JM, Romero M. Effect of firing temperature on sintering of porcelain stoneware tiles. Ceramics International. 2008;**34**(8):1867-1873

[62] El Boudour El Idrissi H, Daoudi L, El Ouahabi M, Collin F, Fagel N. The influence of clay composition and lithology on the industrial potential of earthenware. Construction and Building Materials. 2018;**172**:650-659

[63] Lin T, Zhao Z, Wang T, Pan YT. Three-dimensional printing of large ceramic products and process simulation. Materials. 2023;**16**(10):3815

[64] Rückrich S, Agranati G, Grobman YJ. Earth-based additive manufacturing: A field-oriented methodology for evaluating material printability. Architectural Science Review. 2023;**66**(2):133-143

[65] Mitchell JK, Soga K, O'Sullivan C. Fundamentals of Soil Behavior. Third edition. Hoboken, New Jersey: John Wiley & Sons, Limited; 2024

[66] Carr MM, Wang Y, Ghayoomi M, Newell P. Effects of 3D printing on clay permeability and strength. Transport in Porous Media. 2023;**148**(3):499-518

[67] Chen J, Tong H, Yuan J, Fang Y, Gu R. Permeability prediction model modified on Kozeny-Carman for building foundation of clay soil. Buildings. 2022;**12**(11):1798

[68] Terzaghi K. Principles of Soil Mechanics: A Summary of Experimental Studies of Clay and Sand. New York: McGraw-Hill; 1926. 104 p

[69] Chapuis RP, Aubertin M. On the use of the Kozeny-Carman equation to predict the hydraulic conductivity of soils. Canadian Geotechnical Journal. 2003;**40**(3):616-628

[70] Ji H, Zhao J, Chen J, Shimai S, Zhang J, Liu Y, et al. A novel experimental approach to quantitatively evaluate the printability of inks in 3D printing using two criteria. Additive Manufacturing. 2022;**55**:102846

[71] Boyer SAE, Jandet L, Burr A. 3D-extrusion manufacturing of a kaolinite dough taken in its pristine state. Frontiers in Materials. 2021;**8**:582885. Available from: https://www.frontiersin.org/articles/10.3389/fmats.2021.582885

[72] Alqenaee A, Memari A. Experimental study of 3D printable cob mixtures. Construction and Building Materials. 2022;**324**:126574

[73] Afriat A, Bach JS, Gunduz I, Rhoads JF, Son SF. Comparing the capabilities of vibration-assisted printing (VAP) and direct-write additive manufacturing techniques. International Journal of Advanced Manufacturing Technology. 2022;**121**(11):8231-8241

[74] Gadelmawla ES, Koura MM, Maksoud TMA, Elewa IM, Soliman HH. Roughness parameters. Journal of Materials Processing Technology. 2002;**123**(1):133-145

[75] Mechtcherine V, Bos FP, Perrot A, da Silva WRL, Nerella VN, Fataei S, et al. Extrusion-based additive manufacturing with cement-based materials—Production steps, processes, and their underlying physics: A review. Cement and Concrete Research. 2020;**132**:106037

[76] Mechtcherine V, Nerella VN, Will F, Näther M, Otto J, Krause M. Large-scale digital concrete construction—CONPrint3D concept for on-site, monolithic 3D-printing. Automation in Construction. 2019;**107**:102933

[77] Perrot A, Rangeard D, Courteille E. 3D printing of earth-based materials: Processing aspects. Construction and Building Materials. 2018;**172**:670-676

[78] Jauk J, Gosch L, Vašatko H, Königsberger M, Schlusche J, Stavric M. Filament-reinforced 3D printing of clay. Materials. 2023;**16**(18):6253

[79] Marlin Firmware. Gcode. Marlin Firmware; 2024. Available from: https://marlinfw.org/meta/gcode/

[80] Kim NP, Cho D, Zielewski M. Optimization of 3D printing parameters of screw type extrusion (STE) for ceramics using the Taguchi method. Ceramics International. 2019;**45**(2):2351-2360

[81] Żrodowski Ł, Wróblewski R, Leonowicz M, Morończyk B, Choma T, Ciftci J, et al. How to control the crystallization of metallic glasses during laser powder bed fusion? Towards part-specific 3D printing of in situ composites. Additive Manufacturing. 2023;**76**:103775

[82] Schmidt M, Merklein M, Bourell D, Dimitrov D, Hausotte T, Wegener K, et al. Laser based additive manufacturing in industry and academia. CIRP Annals. 2017;**66**(2):561-583

[83] Djokikj J, Kandikjan T. DfAM: Development of design rules for FFF. Procedia CIRP. 2022;**112**:370-375

[84] Entezari A, Liu NC, Roohani I, Zhang Z, Chen J, Sarrafpour B, et al. On design for additive manufacturing (DAM) parameter and its effects on biomechanical properties of 3D printed ceramic scaffolds. Materials Today Communications. 2020;**23**:101065

[85] Houcine S, Hamid A, Khalid EB. Application of a design for additive manufacturing methodology to optimize the mechanical performance of a PLA test sample. In: Azrar L, Jalid A, Lamouri S, Siadat A, Taha Janan M, Chaari F, et al., editors. Advances in Integrated Design and Production II, Lecture Notes in Mechanical Engineering. Cham: Springer International Publishing; 2023. pp. 490-499

[86] Leary M. Design for Additive Manufacturing. 1st ed. Netherlands: Elsevier; 2019. 348 p

[87] Vaneker T, Bernard A, Moroni G, Gibson I, Zhang Y. Design for additive

manufacturing: Framework and methodology. CIRP Annals. 2020;**69**(2):578-599

[88] Trovato M, Cicconi P. Design tools for metal additive manufacturing: A critical and perspective overview. Procedia CIRP. 2023;**119**:1084-1090

[89] Gunduz IE, McClain MS, Cattani P, Chiu GTC, Rhoads JF, Son SF. 3D printing of extremely viscous materials using ultrasonic vibrations. Additive Manufacturing. 2018;**22**:98-103

Section 2

Technical Studies and Effects on Clay Soil Properties

Chapter 4

The Effect of Samples Disturbance of Partially Saturated Expansive Clay Soils on the Soil Properties

Nariman Hisham Halalo

Abstract

The internal structure of the expansive clay soils is sensitive to the disturbance. When the samples are extracted, the structure of the natural soil is damaged when it is remolded. Therefore, a change in the properties of the expansive soil happens. This matter must be more precisely defined in the design of the foundations of engineering facilities. The objective of this research is to conduct a study of the effect of the disturbance of expansive clay partially saturated soil samples. The obtained results indicate that the behavior of the expansive soil tends to change during the formation of samples. This behavior is based on soil properties, chemical composition and the content of their expansive soil minerals. Soil structure disturbance causes a decrease in soil property variables.

Keywords: expansive soil, partially saturated soil, samples disturbance, soil properties, Direct Shear Test

1. Introduction

An expansive clay soil can be defined as soil whose volume changes with changes in its moisture. The volume of the soil increases when it swells as a result of the increase in the thickness of the water membranes around the soil particles, while when it shrinks, the volume of the soil decreases as a result of the decrease in the thickness of the water membranes around those particles [1].

The behavior of expansive soil is affected by several physical, chemical and environmental factors attributed to the expansive nature of the soil, which is represented by the type and amount of clay minerals, the physical and chemical properties of the expansive clay, soil density, water content, evidence of plasticity, temperature and time [1].

The special behavior of this soil and its direct relationship with water, and the risks that this behavior leads to for various engineering facilities, make this soil a field for much research (**Figure 1**) [2].

Clay minerals greatly affect the volumetric change in expansive soils, as kaolinite, montmorillonite, and Illite are the common clay minerals in expansive soils, and soils containing montmorillonite minerals are swollen in nature [3].

The importance of engineering trial in choosing the distinctive characteristics of expansive soils and the effects of their application and modification in order to ensure

 IntechOpen

Figure 1.
Cracks in the swollen soil of the soil of Umm Rawaq village (Suwayda Governorate – Syria).

the safety and stability of the structure and thus reach the optimal solution for choosing the appropriate type of soil for designing the foundations of engineering facilities [4].

An expansive soil tests have a significant impact on the design of foundations, as they provide sufficient information to determine the characteristics of the expansive soil and its parameters, and to know whether there are technical Remolded that require research and investigation [5, 6].

2. Research justifications

The behavior of expansive soils is a reflection of the capillary property of the soil under the influence of periodic paths of wetting and drying due to natural environmental fluctuations [7]. This behavior is related to the disturbance occurring to the sample when it is transported to the laboratory, which can be classified as follows:

1. Changes in the stress status of the soil.

2. Changes in moisture content and void ratio.

3. Chemical changes.

The change in the stress situation in the soil is related to the water saturation state of the soil, which is either completely saturated with water or partially saturated with water. These saturated states of expansive soil led to thinking that there is a major Remolded in determining the stress state of the soil and the compressibility of expansive soil, especially for large engineering structures such as dams and bridges [8].

The foundations for determining the specifications of the swell rank are based on the results given by intact samples. The Remolded of research is that we are often forced to resort to remolding samples in the laboratory due to our inability to obtain intact samples [9].

In light of this, it is important to recognize indicators of the properties of expansive soils, which are determined by the soil resistance associated with the soil texture, which can be known from Atterberg's limits, especially the liquidity index (LI) [1]. The extent of the soil's ability to swell is related to the plasticity index (PI) and the soil structure, which depends mainly on the mineralogical analysis of the studied samples [10].

3. The main objective of the research

The importance of this research is focused on providing a clear vision of the effect of sample disturbance on the properties of the partially saturated expansive clay soils [11], through:

1. Introducing the concept of disturbance of samples of expansive clay soil in conditions of unstable soil from wetting and drying (partially saturated soil) by presenting

    a. The main causes of sample disturbance.

    b. Effects resulting from sample disturbance.

2. Study the effect of disturbance on the behavior and properties of partially saturated expansive clay soils, which includes physical, mechanical and chemical properties

4. Forms of soil tests

Most tests performed in the laboratory are to determine the physical, mechanical and chemical properties of the soil. The most important of these laboratory and field tests, with reference to the standard references and codes based on which the tests can be conducted. We will mainly refer to the standards of the American Society for Testing and Materials (ASTM) (**Table 1**) [11].

ASTM D-2216	Moisture content (W)
ASTM D-4318	Atterberg Limits
ASTM D-7263	Total unit weight (γ)
ASTM D-854	Specific gravity
ASTM D-422	Sieve analysis
ASTM D-1140	Percent finer sieve 75 micron (sieve # 200)
ASTM D-3080	Direct shear
ASTM D-2435	One-dimensional consolidation
ASTM D-698	Standard Proctor test
ASTM D-5084	Falling Head Soil Permeability Test
Organic fertilizer reference	Organic matter content
ASTM-D 4972	PH value

Table 1.
Laboratory tests according to the American code (ASTM).

5. Samples disturbance

Taking samples is considered one of the most important stages of engineering work, and its importance is no less important than the tests that will be conducted on it. Therefore, it is necessary to exercise accuracy and caution when taking samples and how to fill them so that they are samples representative of the nature of the natural soil. Samples are taken to describe the type of soil and conduct laboratory tests, which are:

1. Classification tests (Atterberg limits).

2. Tests to determine engineering design standards (resistance, compressibility and permeability).

Soil disturbance occurs during excavation, during sampling, during transportation and storage, or during preparation for testing. In general, any soil sample taken from the site and transported to the laboratory will be subject to disturbance. The availability of good engineering standards for design depends on careful testing. The test can be carried out in the laboratory or in the field, but in both cases the most important factor for judging the quality of the final results must be to avoid disturbing the soil samples [12].

6. Main causes of sampling disturbance

They can be classified according to the following table into the various stages of site inspection into [12]:

Reasons before sampling Reasons during sampling Reasons after sampling	Reasons before sampling Reasons during sampling Reasons after sampling	Reasons before sampling Reasons during sampling Reasons after sampling
Reducing or releasing stresses Reducing or releasing stresses. Reducing or releasing stresses	Reducing or releasing stresses Reducing or releasing stresses. Reducing or releasing stresses	Reducing or releasing stresses Reducing or releasing stresses. Reducing or releasing stresses
Swelling Collapse of soil Moisture loss	Swelling Collapse of soil Moisture loss	Swelling Collapse of soil Moisture loss
Pressure displacement effects of sample freezing	Pressure displacement effects of sample freezing	Pressure displacement effects of sample freezing
Displacement cracking or shattering of soil by high temperatures	Displacement cracking or shattering of soil by high temperatures	Displacement cracking or shattering of soil by high temperatures
Foundation stones swell at the cutting edge due to the effects of vibration and shock	Foundation stones swell at the cutting edge due to the effects of vibration and shock	Foundation stones swell at the cutting edge due to the effects of vibration and shock
Collapses, mixing and separation of soil, effects of chemical reactions during pouring	Collapses, mixing and separation of soil, effects of chemical reactions during pouring	Collapses, mixing and separation of soil, effects of chemical reactions during pouring

Table 2.
Main causes of sampling disturbance.

1. Reasons before taking samples.

2. Reasons during sampling.

3. Reasons after sampling (**Table 2**).

7. Effects resulting from sample disturbance

In addition to the process of digging and taking samples, soil disturbance can affect several factors, as the difference in the method of taking samples directly affects the value of the shear resistance of undrained soil, as high-quality methods for extracting samples such as (JPN & Sherbrook) samples, which give undrained shear strength It costs double the values for samples extracted using the (Shelby & NGI54) method (**Figure 2**) [13, 14].

8. Criteria for evaluating the quality of samples

The data values of the studied samples are analyzed to estimate the quality of the sample using several criteria, which is called the sample quality presumption (SQD), defined by the scientist Terzaghi et al. [15] in addition to Lunne et al.'s [16], and these criteria are summarized in the following **Table 3**:

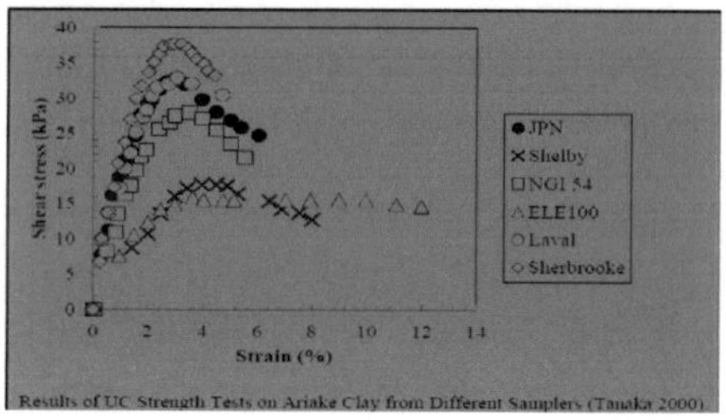

Figure 2.
The effect of the sampling method on the relationship curve between (deformation and shear stress) within the direct shear experiment.

Sample quality standards (SQD)		$\left(\frac{\Delta e}{e_0} = \frac{\Delta V}{V_0}\right)$ *Criteria / Lunne*(*1997*)		
ε_v	SQD	OCR = 1–2	OCR = 2–4	Rating
<1	A	<0.04	<0.03	Very good
1–2	B	0.04–0.07	0.03–0.05	Good to fair
2–4	S	0.07–0.14	0.05–0.10	Poor
4–8	D	>0.14	>0.14	Very Poor
>8	E	—	—	—

Table 3.
Criteria for evaluating the quality of samples according to Terzaghi et al. [15], Lunne et al.'s [16].

9. Precautions to be taken to preserve samples

9.1 Soil sampling tools

There are many tools that are used in taking soil samples, such as a soil roller, auger, shovel, soil auger, drill, and rings for taking soil samples (**Figure 3**).

9.2 How to take soil samples from the field and prepare them for analysis

A field visit is made to the field or study area and some information is recorded about it by looking at it, such as the topography of the land, the condition of the plants there (if any), construction, texture, and the condition of the ground water. Then a sketch drawing of the field is made and the fixed places surrounding the land are placed on this drawing (Roads, mosques and churches) He also signs the points from which samples will be taken. After completing the previous step, he goes down to the field (study area) and takes samples from the specified places using the appropriate tool.

After taking the sample, it is placed in bags bearing the sample data (such as the sample number, sample coordinates, and the depth from which the sample was taken). The sample is then transported to the laboratory and spread out to air dry at the temperature of the laboratory. After that, we keep it in the bags with the data written on it until it is analyzed.

9.3 Laboratory description of the soil

We have mentioned that the soil must be described by mentioning its necessary engineering properties. Therefore, the size of the soil grains constitutes an essential factor in describing non-cohesive soils, while cohesive (silty) soils are described according to their degree of cohesion, as the percentage of clay present in the soil constitutes the controlling element on the soil's properties and behavior.

One of the difficulties encountered in describing soil is that most soils are a mixture of gravel, sand, silt, and clay, and there are general methods that must be followed when we want to describe a soil in the laboratory.

1. We read the description page provided by the sounding controller and compare its numbers with the numbers of the samples to be described, then we open the boxes carefully and with special care if the samples are intact.

Figure 3.
Taking a soil sample using a shovel.

2. We arrange the samples logically according to their numbers so that we separate healthy samples from damaged samples.

3. We remove the wax cover with a sharp cutting tool so that the sample is not exposed to damage or scratches.

4. We must maintain the degree of humidity of the sample as much as possible and not remove wax from sealed samples except at the time of use, and always take only the necessary amount. We also try to prevent changing its chemical composition due to oxidation in the air.

5. We examine the sample without destroying it, and we specifically mention the degree of its cohesion or disintegration [17].

9.4 Sample filling

Immediately upon obtaining the samples, the samples are filled in sealed containers, such as plastic containers or plastic bags, and then placed inside fabric bags, taking care not to crush them when inserting them into the bag. The sample must fill the container as much as possible, and in the event that the sample is a continuous sample Like rock samples, they are kept in boxes with divisions of appropriate diameters so that they hold the samples without compressing them. However, in the case of extracting healthy samples, these samples must be protected in appropriate ways from drying out, changing their size, or slipping in the container.

As for samples taken from cohesive soil and cut into cubes, the samples can be covered well with one or more layers of wax, and each sample is placed separately in an outer cover with the same dimensions of wood or something similar to protect it during transportation.

The sample is usually coated with wax after melting it to maintain its moisture and structure by [17]:

a. Paraffin wax.

b. Labels that are attached or affixed to the sample or the container that contains it.

c. Sequentially numbered stickers.

d. Electric heater to heat the wax.

e. Containers for storing extracted samples.

f. Samples must be documented according to their number in the form of a booklet or a packet of papers.

g. Racks for storing samples.

9.5 Waxing samples

Its purpose is to preserve the moisture of the samples in their natural state and to preserve their structure from damage.

Intact block or cubic samples can be waxed by pouring wax on them, with the advisability of covering them once or twice with a preliminary coating using a brush so that the wax does not enter into their cracks and pores. Then the entire sample can be dipped in molten wax, and the work is repeated until a wax layer of about (2) thickness is obtained. 5 mm), and to protect the sample from shattering, the sample can be wrapped in a linen cloth after covering it with wax, then immerse it again with the cloth in the molten wax (**Figure 4**) [17].

9.6 Storage of samples

Samples are stored in sealed containers and placed on exposed shelves. Healthy samples must be separated from spoiled samples. The destroyed samples are stored in bottles or metal containers of an appropriate size in order to reduce the loss of moisture as much as possible. Cards are placed inside the containers and numbered stickers are affixed to them on the outside of the containers or bottles. Then a list of the samples, their type and numbers is written, mentioning the date the sample was stored if we want to store a large number of samples. Samples for long periods of time. In all cases, the following data must be recorded when sampling [12]:

a. The general location, with its clarification on a sketch plan.

b. General information about the project.

c. Hole number and dimensions.

d. Number of samples and locations of extraction.

e. Date the sample was taken and weather conditions.

f. Sampling method.

g. Approximate area or quantity.

h. Groundwater level, if detected.

i. General description of the soil.

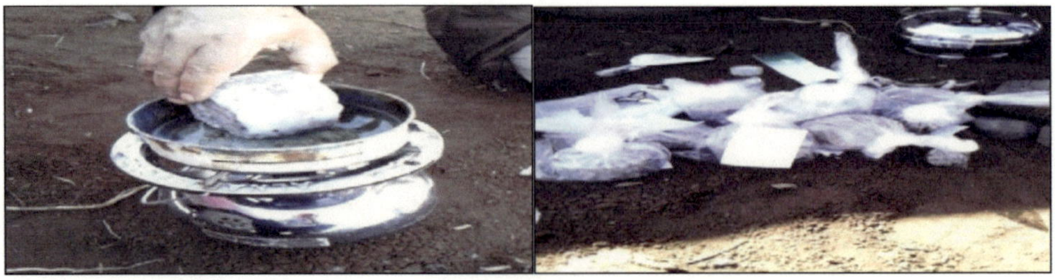

Figure 4.
Packaging and waxing of samples used in research.

9.7 The importance of sample formation

Sample formation is of utmost importance in geotechnical science because it has a clear impact on laboratory results and its reflection on the design reality later. This is due to our inability to obtain intact samples during drilling (the difficulty of drilling, the nature of the site, the use of water during drilling) and the inability to obtain samples. Sufficient in length, therefore, the sample must be reconstituted to understand the behavior of the soil and determine its variables. The destroyed sample can be utilized by classifying the soil and determining the classified physical specifications (Atterberg limits, granular sieve analysis).

The basic physical specifications (natural humidity, normal volumetric weight, dry volumetric weight) can be determined if there are pieces of the destroyed sample that are similar to the mother soil (do not use water during excavation) and the mechanical specifications (cohesion and friction) cannot be determined, so the sample must be formed according to one of the codes. Then, experiments were conducted on it, including direct shear experiments and compression experiments, in order to determine the mechanical specifications and benefit from them in determining the bearing capacity and subsidence [18].

9.8 Methods used in sample formation

There are several ways to form samples according to the desired purpose of forming the sample to serve the desired scientific purpose, and methods of forming samples according to the American Code (ASTM).

Samples are prepared using the specified optimum moisture content and maximum dry volumetric weight. The sample is formed either by compacting it or knocking it into layers. The amount of soil for each layer is determined and placed in the device. Then the sample is compacted to achieve the required volumetric weight. The soil continues to be added and compacted until the entire sample is compacted. The materials subjected to the test will be completely mixed with a sufficient amount of water to obtain the desired moisture content, and the samples will be left for periods of time according to the classification (ML, CH). According to the instructions and according to the American Code, the minimum waiting time is (18 hours).

- The destroyed sample is formed [19] and requires passing the soil through sieve No. (#4) and drying it in the oven at a temperature of (105°–110°C). The sample similar to the original sample is formed by hammering it into the mold (direct shear box – odometer ring). Within three layers, each layer with (25 strokes), thus we obtained a sample similar to the sample formed according to the regular Proctor experiment.

- The formed sample is formed, which is obtained by the regular Proctor experiment (optimal humidity and maximum dry volumetric weight), where the first and last 3 cm are scraped off from the Proctor mold sample so that the middle part remains, and from it a sample is taken for the direct shear square and a sample for the odometer ring.

Mold specifications for the regular Proctor experiment (D = 10.2 cm, h = 11.6 cm). Where h: height of the cylinder, D: diameter of the cylinder.

Number of layers: Three layers, with (25) strokes for each layer.

Hammer Weight: (2.5 kg), Screen No.: #4), Drop Height: 12″).

Therefore, all damaged and damaged samples have the same ideal humidity and the same volumetric weight, and they can be taken as a group and one spirit for study, and thus the effect is the same on all samples.

10. Reference studies

There are many previous studies that addressed the subject of studying the effect of disturbance on swollen soil samples partially saturated with water and approached it from different angles.

10.1 Search [A1]

The researcher [20] studied the effect of sample disturbance on the behavior of moderately plastic clay soil. He conducted tests on vandalized samples and compared them with intact samples within the odometer experiment (CAUC-L). The samples were loaded with increasing loads (0.25 – 0.5 – 0.75). – 1 kg/cm^2) for (24) hours for each stage, and in order to obtain values that mimic reality, standards such as:

Types of samples: (ratio of sample height to diameter for all samples (h/d = 1.8).

1. An intact sample using a tube with a cutting edge (tube sample 30°) and a diameter of 100 mm.

2. An intact sample using a tube with a cutting edge (tube sample 5°) and a diameter of 100 mm.

3. An intact sample using a block sample tool with a diameter of 50 mm.

4. A vandalized sample (reconstructed) (Mostap sample) with a diameter of 50 mm (**Figure 5** and **Table 4**).

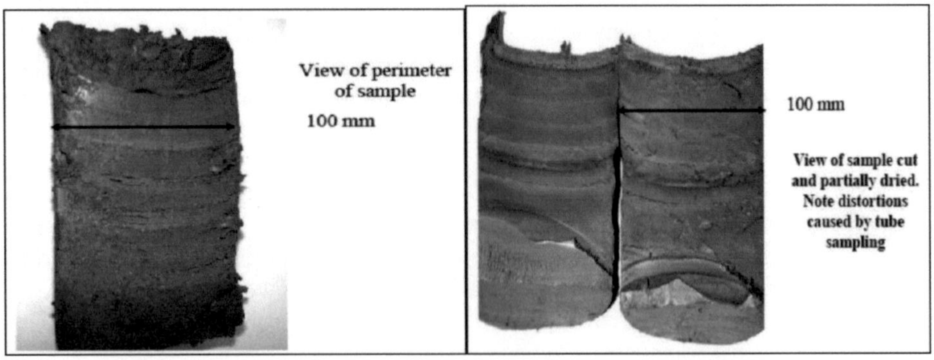

Figure 5.
Distortions caused by a sampling tube for a dried soil sample [20].

Percentage of volumetric deformation for each sample	
Intact sample (tube 30°)	εv = 5.9%
Intact sample (tube 5°)	εv = 4%
Intact sample (block)	εv = 4.3%
Disturbance sample	εv = 7.8%

Table 4.
Mechanical properties of moderately plastic clay soil samples [20].

According to the results of the study, the research concluded that the results of sample quality are related to the percentage of volumetric deformation, and the relationship between the variables was that the sample was intact (tube 5°) the best sample (**Figure 6**).

10.2 Search [A2]

Amundsen et al. [21] studied the effect of disturbance of intact samples (block samples) on the behavior of low-plastic clay soil by studying the relationship curve (horizontal deformation – shear stress) within a triaxial experiment And the odometer test (CAUC triaxial and CRS Oedometer tests), as shown in **Figure 7**, for both intact, low-plasticity swelling soil samples that were taken using a block sample method with dimensions (160 × 160 mm) and other samples that were taken using a diameter tube (75 mm).

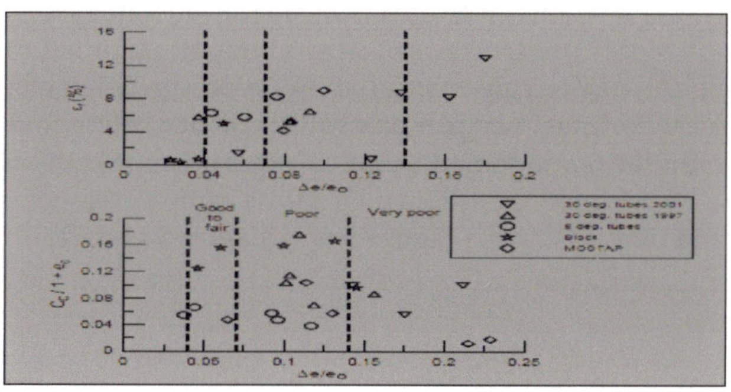

Figure 6.
Comparison of the various results of the odometer experiment for good quality values [20].

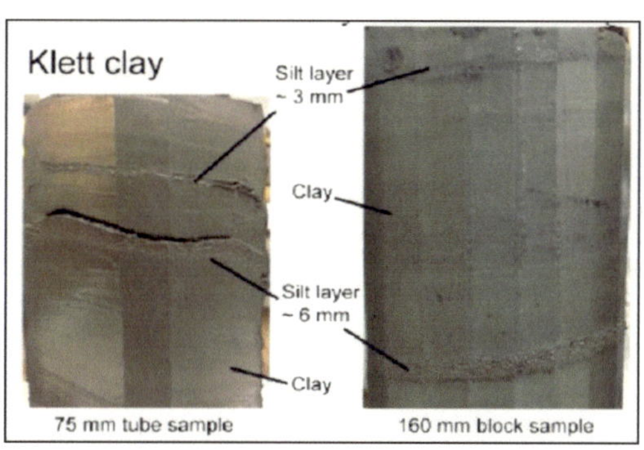

Figure 7.
Block samples and tubular samples [21].

According to the results of the study, the research concluded that the shear strength and specifications of block specimens are generally better than tubular specimens, and the decrease in the strength of tubular specimens is due to the destruction of structural bonds in the soil (**Figure 8**).

10.3 Search [A3]

Al-Miqdad studied the effect of sample reshaping on mechanical specifications in clay soil by conducting laboratory experiments on intact samples and formed samples, which are similar to the intact sample in terms of moisture and natural volumetric weight, within a direct shearing experiment (fast – undrained) at a cutting speed (0.5 mm/min).), the applied stresses (1.11 – 2.22 – 3.33 kg/cm^2) Several sites were chosen for the study (Jdeidet al Wadi – Deir ez-Zor – Sahnaya – Ghabagheb – Dabbagat). The results of the study were as follows [17]:

1. The relative change in soil cohesion between the healthy sample and the Remolded sample ranged between (8–56) %.

2. The relative change in the angle of internal friction of the soil between the intact sample and the Remolded sample ranged between (3–9) %.

10.4 Search [A4]

The researcher [22] studied the effect of the degree of water saturation of the soil on the behavior of a highly plastic clay soil by studying the size and rate of subsidence of the swollen soil before construction in a certain condition of soil moisture, by studying the behavior of intact samples of swollen soil according to different degrees of saturation. Within the odometer and triaxial experiment (CAUC-UL) [23].

The samples used are intact samples using a tube sample, diameter (75 mm), height (40 mm) and depth (1.5 m) (**Figure 9** and **Tables 5** and **6**).

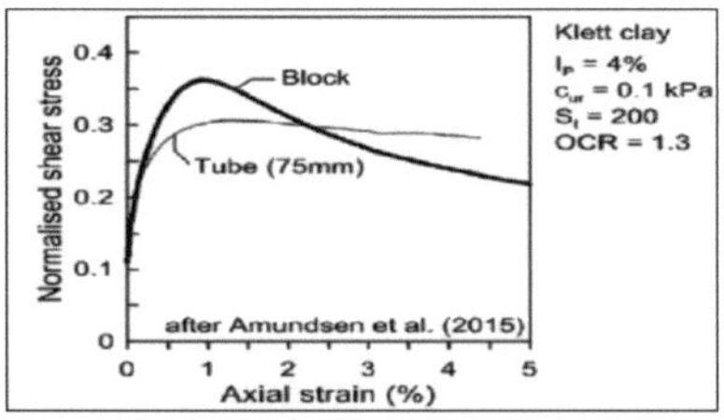

Figure 8.
Relationship curve (horizontal deformation – shear stress) [21].

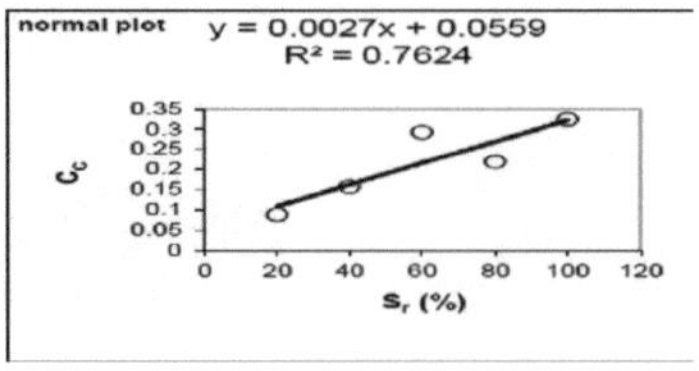

Figure 9.
The relationship between the compressibility index and the degree of water saturation of the soil [22].

Cartilage activity	1.73
The type of metal present	Montmorillonite

Table 5.
Chemical properties of highly plastic clay soil samples [22].

τ (kN/m²)	Cc	W (%)	Sr (%)
491.4	0.088	1.85	20
485.8	0.157	3.69	40
362.3	0.2917	5.54	60
243.7	0.2181	7.38	80
209.3	0.3238	9.23	100

Table 6.
Mechanical properties of highly plastic clay soil samples [22].

According to the results of the study, the research concluded that there is a linear relationship between the initial compressibility index and the degree of water saturation of the soil. As the degree of water saturation of the soil increases, the compressibility index increases, and this relationship is given in the following form: Cc = 0.0027*Sr. + 0.0559.

10.5 Search [A5]

The researcher [24] studied the change in moisture content with changes in pore pressure in the behavior of swelling soil, where a comparison was made between intact soil samples within the odometer experiment (CAUC-UL) [24].

The samples used are intact samples using a tube sample with a diameter of 6.35 cm and a height of 1.7 cm. Sample depths: from 0.5 m to 1.00 m (**Figures 10** and **11** and **Tables 7–9**).

The research concluded according to the results of the study:

1. Swelling of soil is related to the properties of the soil, its chemical composition, and the content of metals in the soil.

2. The behavior of swelling soil is related to the change in water content in the soil according to a linear relationship.

3. The degree of soil saturation with water is linearly related to the percentage of soil deformation, and swelling stops at a total degree of saturation.

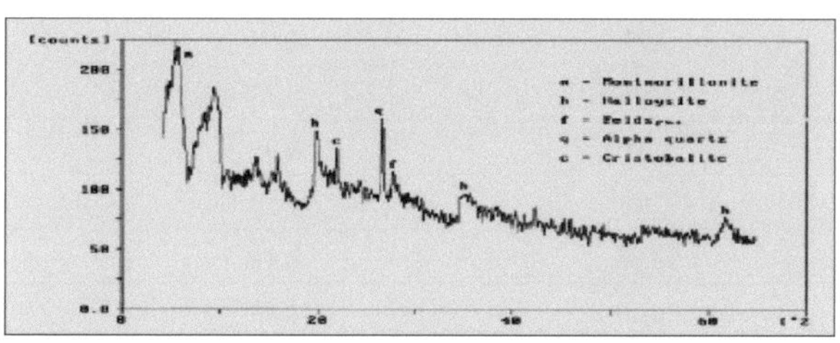

Figure 10.
Mineral analysis of the soil sample (X-ray diffraction of clay Soko, Ngawi).

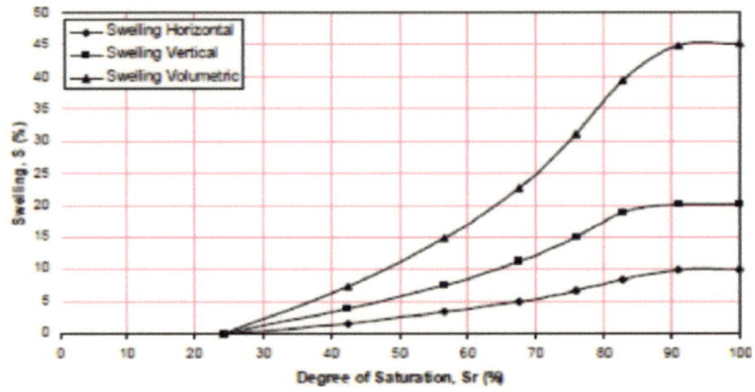

Figure 11.
The relationship between the degree of water saturation and the percentage of vertical, horizontal and volumetric swelling of the soil [24].

Relative composition (%)	Composition type
76.1	SiO_2
18.59	Al_2O_3
2.75	Fe_2O_3
1.80	CaO
0.50	MgO
0.22	Na_2O
0.04	K2O

Table 7.
Chemical properties of intact clay soil samples [24].

Relative composition (%)	Type of minerals
76.1	Montmorillonite
16.20	Feldspar
5.3	Alpha Quartz
4.30	Halloysite
0.90	Cristobalite

Table 8.
Relative composition of clay soil minerals [24].

Sr (%)	W (%)	Vertical swelling Sx (%)	Horizontal swelling Sy (%)	Volumetric swelling Sv (%)
24.27	10	0.0	0.0	0.0
42.49	20	3.99	1.62	7.38
56.61	30	7.54	3.34	14.85
67.54	40	1.31	4.99	22.70
75.88	50	1.09	6.73	31.10
82.81	60	18.86	8.44	39.37
91.10	70	20.10	9.83	44.89

Table 9.
Mechanical properties of intact clay soil samples [24].

11. Research materials and methods

11.1 Swelling soil

Soil samples were brought from the Damascus countryside governorate and As-Suwayda governorate. The samples were extracted using a simple drilling mechanism to a depth of (60 cm) from the bottom of the soil layer. The samples were brought in sealed plastic bags, and then two samples were studied for each site and the samples were named according to the following **Table 10**.

11.2 Method of forming samples

The method is based on conducting all experiments on the basic physical, classified, mathematical, mechanical and chemical properties of the soil used in this research [4], taking into account the conditions stated in the American specifications (ASTM). In the second stage, the shear resistance properties of the studied soil were calculated (angle of internal friction, cohesion Soil), [14] within a direct shearing experiment [11] for vandalized soil samples and comparing them with samples formed in the laboratory.

- Testing within a direct shear experiment on a sample (uncompressed – undrained),
- (Unconsolidated Undrained Direct Shear Test)
- The sample inside the direct shear device cell was subjected to shear stress at a constant speed (0.2 mm/min).
- Internal shear box dimensions (6.1 cm × 6.1 cm), shear box height (2.3 cm).

The disturbance sample was formed according to the American specification (ASTM D-3080) [11], which requires passing the soil through sieve No. (4) and drying it in the oven at a temperature of (105–110°C). The sample similar to the original sample is formed by pounding it into the mold. (Direct shearing box) in three layers, where each layer has (25 strokes), and thus we obtained a sample similar to the sample formed according to the regular proctor.

The formed sample was formed according to the American standard (ASTM D-3080), which is obtained using a regular proctor (optimal humidity and maximum dry volumetric weight), where the first and last 3 cm of the proctor mold sample were scraped off so that the middle part remained, and from it a sample was taken for the shear square. Direct.

Site code	Location	Soil color
A	Umm Rawaq Village – Suwayda	dark brown
B	Deir Al-Hajar village – Al-Ghazlaniyah – Damascus countryside	light brown

Table 10.
Soil sample locations.

ASTM	A	B	Units
Specific gravity of solids, Gs	2.68	2.71	g/cm^3
Optimum moisture content, wi	24.30	20	%
Maximum Dry density, w Optimum	1.42	1.55	g/cm^3
Degree of saturation, Sr	72.85	71.88	%

Table 11.
Results of water saturation values for soil samples within the direct shear experiment.

- Thus, the American standard (ASTM D-3080) completely regulates the method of forming samples, and therefore all spoiled and formed samples have the same ideal humidity and the same volumetric weight, and therefore they can be taken as a group and one spirit for study, and the effect is the same on all samples.

- The degree of water saturation of the soil samples from the direct shear experiment was determined by knowing the specific gravity, optimum moisture, and dry volumetric weight according to the Proctor experiment of the soil sample through the following relationship:

$$S_r = \frac{w^* G_s}{e} = \frac{w^* G_s}{\frac{G_s^* \gamma_w}{\gamma_d} - 1} \quad (1)$$

S_r : Degree of water saturation (%)
G_s : Relative specific gravity
γ_d : Dry volumetric weight of soil g/cm^3
γ_w : The volumetric weight of one unit of water, g/cm^3
w : Optimum humidity according to Proctor's systematic experience
e : Coefficient of porosity
See **Table 11.**

12. Results and discussion

12.1 Results of physical properties

12.1.1 Gradient

The particle size of each of the studied soils was determined by the granular sieve analysis method for particles with diameters larger than (0.075 mm), according to (ASTM-D422), and by the hydrometer sedimentation method for particles with diameters less than (0.075 mm), according to (ASTM-D1140) (**Figure 12**).

12.1.2 Basic physical properties

See **Table 12.**

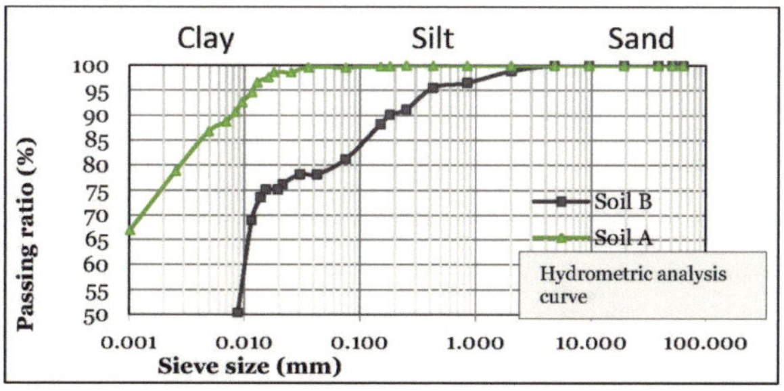

Figure 12.
Hydrometric analysis curve.

ASTM	Site		Unit
	A	B	
Natural moisture content	37.1	23.57	%
Relative specific gravity	2.68	2.71	g/cm^3
Volumetric weight	1.62	1.57	g/cm^3
Dry volumetric weight	1.18	1.27	g/cm^3

Table 12.
Results of basic soil physical properties experiments.

12.1.3 Computational physical properties

See **Table 13**.

12.1.4 Atterberg limits results

See **Table 14**.

12.1.5 Classification of the studied soils

The soils for the studied samples were classified according to the American Standard Soil Classification System (USCS) using the Casagrande plasticity scheme [5] as follows:

1. Highly plastic clay soil (CH), which is the soil (Umm Rwaq village), site (A).

2. Low plasticity (ML) silty soil (Deir Al-Hajar village), site (B) (**Figure 13**).

ASTM	Site		Unit
	A	B	
Voids ratio	1.27	1.13	%
porosity	0.56	0.53	g/cm^3

Table 13.
Results of computational physical properties experiments.

ASTM	Site		Unit
	A	B	
Liquidity limit	65.92	42.73	%
Plasticity limit	33.33	36.76	%
Liquidity index	0.12	0.48	_
Plasticity index	32.59	5.97	%
Percentage of fineness (passing through sieve #200)	97.54	78.18	%
Gluten content	74.56	33.88	%
Celt content	25.44	47.12	%
Sand content	2.46	19	%

Table 14.
Results of experiments on classified soil physical properties.

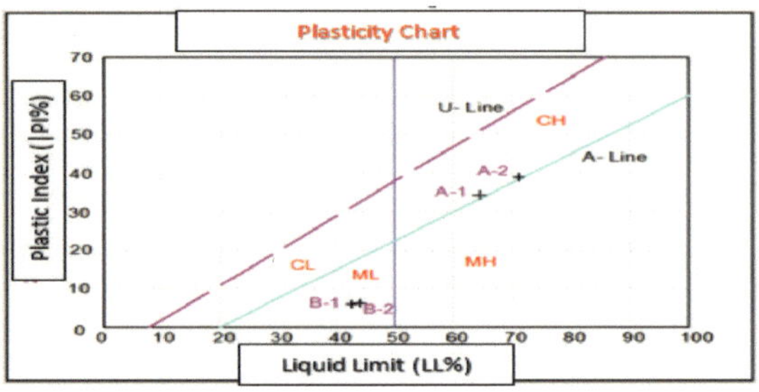

Figure 13.
Soil classification scheme according to the plasticity scheme.

12.1.6 Results of the systematic Proctor experiment

The aim of conducting the compaction experiment is to determine the maximum dry density and ideal moisture of the studied samples in order to form the samples according to the American standard (ASTM D-3080) (**Figure 14** and **Table 15**).

12.1.7 Results of chemical experiments

See **Table 16**.

Figures 15 and **16** represent the metallic analysis of samples using X-ray diffraction, which was conducted at the General Institution of Geology [19].

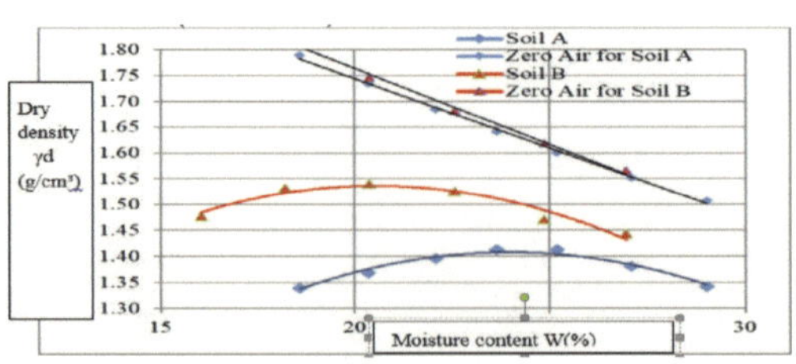

Figure 14.
Compaction curves for the soil of the two sites (A–B).

ASTM	Site		Unit
	A	B	
Ideal humidity	24.3	20.0	%
Maximum dry density	1.42	1.55	g/cm^3
Saturation humidity	33.4	27.8	%

Table 15.
Results of the systemic Proctor experiment.

ASTM	Site		Unit	Comments
	A	B		
Organic matter content	24.3	20.0	%	It is greater than (2%) and the soil is rich in organic matter
Determine the pH	1.42	1.55	g/cm^3	The soil is moderately acidic

Table 16.
Results of the chemical properties of the soil used in the research.

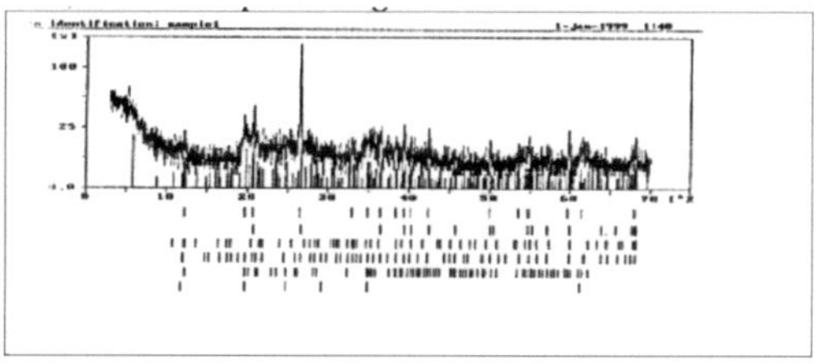

Figure 15.
Mineral analysis of soil sample /A/.

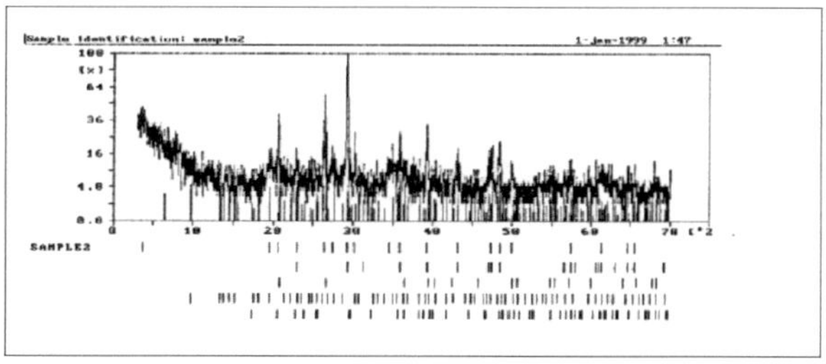

Figure 16.
Mineral analysis of soil sample /B/.

The most abundant clay mineral in both sites is quartz, which is a mineral that has a weak relationship with water. The presence of quartz in high proportions in the two types of soil explains the low values of plasticity, especially since quartz is known to prevent sintering of soil particles and thus gives low plasticity [25], as in the soil of the site. (B). The soil samples from site (A) contain swollen clay minerals such as montmorillonite and kaolinite, but in low percentages (**Table 17**).

Types of metals found in the soil	Sample
Quartz – Phillipsite – Gismodine – Kaolinite-Montmorillonite	A
Calcite – Quartz – Mordenite – Forsterite	B

Table 17.
Types of metals found in the samples used in the research.

12.2 Results of direct shear experiments

The figures show the horizontal deformations recorded for the studied specimens with shear stress plotted in normal coordinates.

Highly plastic soil (fertilized soil-A) (**Figure 17**):
Low plasticity soil (fertilized soil-B) (**Figure 18**):
Highly plastic soil (Remolded soil – A) (**Figure 19**):
Low plasticity soil (B-Remolded soil) (**Figure 20**):

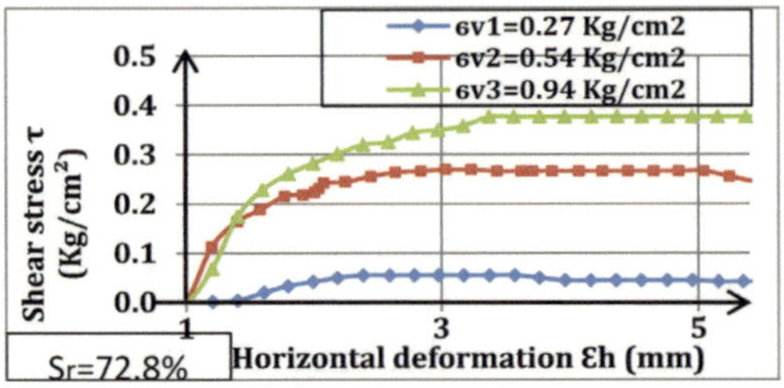

Figure 17.
The relationship curve between (horizontal deformation and stress) for vandalized samples (A).

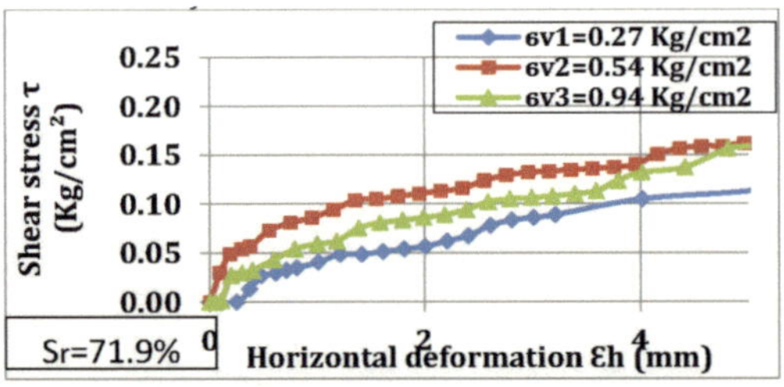

Figure 18.
Relationship curve between (horizontal deformation and stress) for vandalized samples (B).

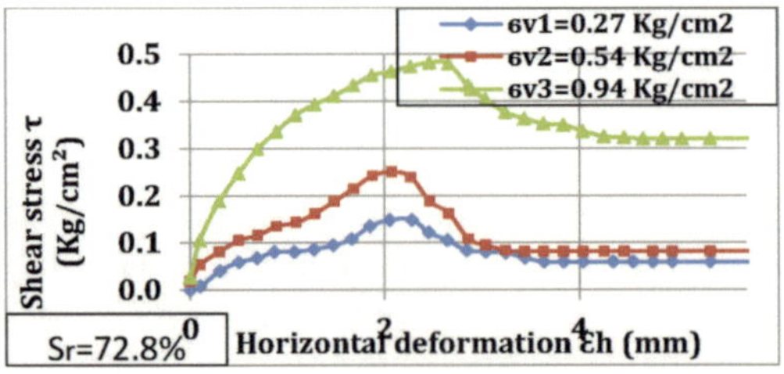

Figure 19.
Relationship curve between (horizontal deformation and stress) for formed samples – (A).

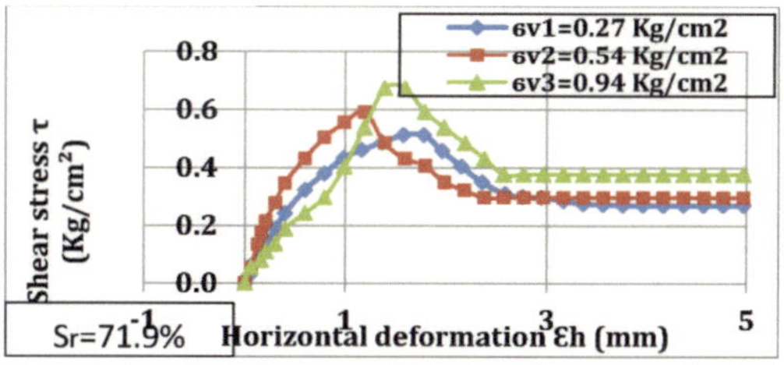

Figure 20.
Relationship curve between (horizontal deformation – stress) for formed samples – (B).

13. Conclusions

Since determining the behavior of swelling soil depends mainly on the metallurgical analysis of the samples, and by comparing the curves of the relationship between (horizontal deformation – shear stress), it was noted that the effect of the disturbance of the samples on low plasticity soil is greater compared to high plasticity soil, as the results of the metallurgical analysis of the samples using X-rays showed X-ray diffraction The types of metals present in the soil samples used in this research are the dominant minerals montmorillonite and kaolinite in terms of their effect on the swelling behavior of the site's soil (A (high plasticity soil), which is characterized by its high ability to absorb water [26]. It has the advantage of cohesion, which helps maintain the shape of the damaged or laboratory-formed clay sample, thus reducing the amount of disturbance occurring to the sample.

14. Results

Through the previous curves, the results were analyzed to find the amount of relative change between the following variables:

a. The amount of relative change in soil cohesion for vandalized samples and laboratory formed samples.

b. The amount of relative change in the angle of internal friction of the soil for vandalized samples and laboratory-formed samples.

Highly plastic soil (**Figure 21**):
Low plasticity soil (**Figure 22**):

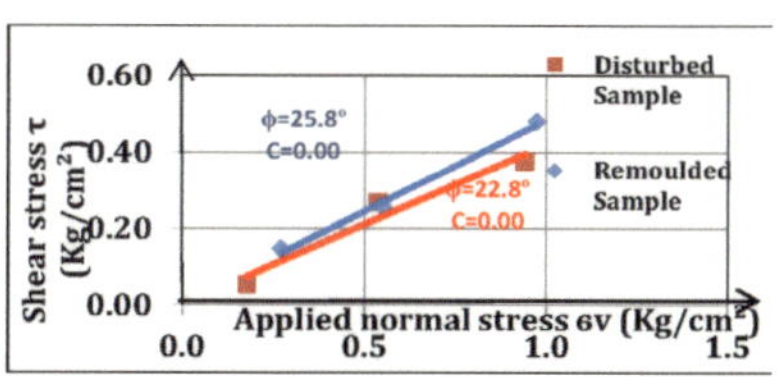

Figure 21.
Coulomb's straight line for damaged samples with shaped samples (A).

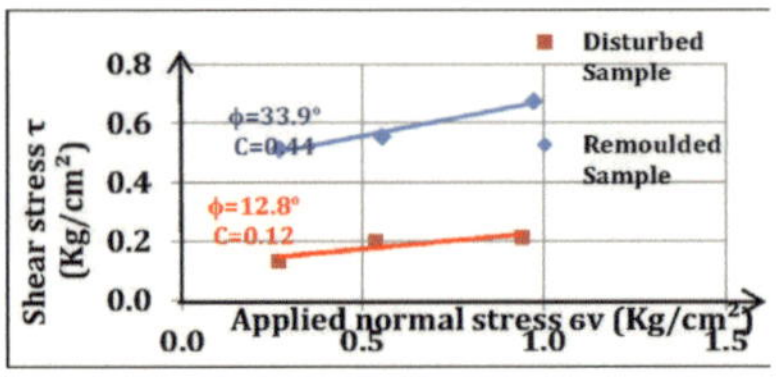

Figure 22.
Coulomb's line for damaged samples with formed samples (B).

$(\Delta\phi/\phi)$ (%)	$\Delta C/C$ (%)	Remolded sample		Disturbance sample		Site
		ϕ (°)	C (kg/cm^2)	ϕ (°)	C (kg/cm^2)	
10.7	75.0	26	0.00	23	0.01	A
62.2	60.0	34	0.44	13	0.12	B

Table 18.
Results of the relative change in the mechanical specifications of the studied soil samples.

The previous results match what was reached in similar studies, where it was found that the disturbance occurring in the vandalized samples leads to a decrease in the values of both soil cohesion and the internal friction angle of the bulging soil, but the amount of decrease was greater for the low-plasticity soil than for the high-plasticity soil, as shown in the following **Table 18**:

After analyzing the results, we reached in the previous paragraphs, we reach the following final conclusions and recommendations in this research:

- The behavior of swelling soils tends to change during the formation of samples, and this behavior depends on the properties of the soil, its chemical composition, and the content of swelling metals in the soil.

- Disturbance of the soil structure causes a decrease in the shear resistance variables of the soil (angle of internal friction, soil cohesion), as the disturbance occurring in the vandalized samples leads to a decrease in the values of both soil cohesion and the angle of internal friction for the swelling soil, but the amount of decrease for low-plasticity soil was greater than for high-plasticity soil. Plasticity.

- The amount of relative change in soil cohesion between the destroyed sample and the formed sample ranged between ($\Delta C/C$ = 60–85%).

- The amount of relative change in the angle of internal friction of the soil between the damaged sample and the formed sample ranged between.

- Since laboratory evaluation of soil specifications is carried out by conducting laboratory experiments on intact samples, and due to our inability to obtain intact samples, we are often forced to reshape the samples in the laboratory, and therefore the possibility of obtaining mechanical soil specifications (angle of internal friction, soil cohesion) is achieved. Through samples formed according to the regular proctor according to the American code.

Author details

Nariman Hisham Halalo
Civil Engineering Faculty, Department of Geotechnical Engineering, Damascus University, Damascus, Syrian Arab Republic

*Address all correspondence to: nar-119@hotmail.com

References

[1] Yeni H. Soil Mechanics. Syria: Damascus University Publications; 1982

[2] Lambe TW, Whitman RV. Soil Mechanics. New York, NY, USA: Wiley; 1969

[3] Gillot E. Clay in Engineering Geology. Amsterdam: Elsevier Publishing Company; 1968. p. 234

[4] Budhu M. Soil Mechanics & Foundations. New York, NY, USA: Wiley; 2000

[5] Escario S, Saez J. The shear strength of partly saturated soils. Geotechnique. 1986;**36**(4):453-456

[6] Fredlund DG, Rahardjo H. Soil Mechanics for Unsaturated Soils. New York: John Wiley & Sons; 1993

[7] Das BM. Principles of Geotechnical Engineering. 4th ed. Boston, MA, USA: PWS Publishing Company; 1998

[8] Caruso A, Tarantino A. A shear box for testing unsaturated soils from medium to high degree of saturation. Geotechnique. 2004;**54**(4):281-284

[9] Snethen DR. Evaluation of expedient methods for identification and classification of potentially expansive soils. In: Proceedings of Fifth International Conference on Expansive Soils. Adelaide, South Australia; 1984. pp. 22-27

[10] Al-Qasabi AF. Soil Mechanics. Syria: Dar Al-Kutub Al-Ilmiyyah for Publishing and Distribution; 1999

[11] ASTM – American Society for Testing and Materials. The American Society for Testing and Materials (ASTM) is a nonprofit organization in which producers, users, consumers, and representatives of government and academia develop voluntary consensus standards for materials, products, systems, and services. ASTM has 30,000 individual and organizational members worldwide

[12] Clayton CRI, Matthews MC, Simons NE. Site investigation. In: Proceedings of the Institution of Civil Engineers – Geotechnical Engineering. 1999

[13] Tanaka H. Sample quality of cohesive soils: Lessons from three sites, Ariake, Bothkennar and Drammen. Soils and Foundations. 2000;**20**(4):57-74

[14] Tanaka H, Sharma P, Tsuchida T, Tanaka M. Comparative study on sample quality using several types of samplers. Soils and Foundations. 1996;**36**(1):57-68

[15] Terzaghi K, Peck RB, Mesri G. Soil Mechanics in Engineering Practice. New York: John Wiley & Sons Inc.; 1996

[16] Lunne T, Berre T, Strandvik S. Sample disturbance effects in soft low plastic Norwegian clay. In: Symposium on Recent Developments in Soil and Pavement Mechanics. 1997. pp. 81-103

[17] Al-Miqdad F. The Effect of Sample Reshaping on the Mechanical Specifications in Clay Soil. Syria: A Study Prepared to Obtain a Master's Degree in Civil Engineering – Department of Geotechnical Engineering – University of Damascus; 2009

[18] Ladd CC. Settlement Analysis of Cohesive Soils. Cambridge, MA: MIT; 1971

[19] ASTM – American Society for Testing and Materials. ASTM-D3080 - Standard Test Method for Direct Shear Test of Soils Under Consolidated Drained Conditions 1. 1989

[20] Long M. Sampling disturbance effects on medium plasticity clay/silt. Journal of Geotechnical and Geoenvironmental Engineering. 2006;**159**(2):99-111

[21] Amundsen HA, Thakur V, Emdal A. Sample Disturbances in Block Samples on Low Plastic Soft Clays. Queensland, Australia: Norwegian University of Science and Technology; 2016

[22] Ojuri OO. Effect of Soil Suction on the Compressibility and Strength of North Eastern Nigerian Black Cotton Soil. Akure, Ondo State, Nigeria: Department of Civil Engineering, Federal University of Technology; 2012

[23] Skempton AW. The colloidal activity of clays. In: Proceedings of the 3rd International Conference on Soil and Mechanics Foundation Engineering. Zurich, Switzerland; 1953. pp. 57-61

[24] Agus TS, Kabul BS, Ahmad R, Indrasurya BM. The effect of water content change and variation suction in behavior swelling of expansive soil. International Journal of Civil & Environmental Engineering IJCEE-IJENS. 2011;**11**(3):14-16

[25] Gan JKM, Fredlund DG, Rahardjo H. Determination of the shear strength parameters of an unsaturated soil using the direct shear test. Canadian Geotechnical Journal. 1988;**25**(8):500-510

[26] Chen FH. Foundations on Expansive Soils. Amsterdam: Elsevier Scientific Publishing Company; 1988

Chapter 5

Effect of Compaction Energy on the Behavior of Coefficient of Consolidation for the Compacted Fine-Grained Soils

Heggadadevanakote Subbarao Prasanna and Unnam Anil

Abstract

Construction exploits marginal areas and landfills due to a lack of suitable worksites. Several researchers are interested in the engineering behavior of compacted fine-grained soils. Clay mineral composition suggests compacted fine-grained soils behave physicochemically. Six natural soils and one artificial soil with varying clay mineralogical compositions (kaolinite, montmorillonite, and K-M) and liquid limits (46, 55, and 68%) were selected and they were categorized into three different groups [G-1 (46%) {K & M-soils}, G-2 (55%) {K & M-soils}, and G-3 (68%) {K, M, & K-M soils}. Consolidation tests were conducted in one dimension under various placement conditions (95% of optimal on dry and wet sides and at optimum), energy levels (Light (LC) and Heavy Compaction (HC)), and seated pressures (σ) from 6.25 to 1600 kPa (@ load increment ratio of 1). The effect of energy level concept was studied by defining the energy ratio of Cv (CvER) = {Cv @ HC/Cv @ LC}) and average ratio (CvAR = Average of CvER) using Proctor compaction energy ratio or standard energy ratio (SER). The values of Cv for heavy compaction can be estimated directly from light compaction energy level values using correlations (R^2 = 0.86 to 0.99).

Keywords: compacted fine-grained soils, clay mineralogy, energy ratio, pressure, placement condition

1. Introduction

There are three primary criteria of structures found on soils: strength, stiffness, and stability, for which the subsoil is expected to be in a compacted state. These properties provide the scope for numerous studies on compacted soils. The study of compacted soils becomes all the more critical in the present-day scenario wherein the lack of good bearing capacity in constructional sites is forcing the people to use the sites that have been considered for ages as unsuitable for constructional activities. Compressibility is one of the essential factors to be considered, and it is because the number of good construction sites available for construction is becoming a significant challenge for the construction industry. It is imperative that the construction be done on soils with low bearing capacity, susceptible to large settlements. Given the ever-growing demand for construction sites, it is inevitable to reclaim the marginal lands after subjecting them to various ground improvement techniques. In the consolidation behavior of the soil, the main factor controlling the soil behavior like settlement and time of consolidation is the Coefficient of Consolidation (Cv). The Cv depends upon the soil's compression and void ratio and the load or pressure acting on it. The compression of soil depends upon the clay mineralogical composition of the soil. In this present experimental study, the variation of Cv is compared for the soils compacted at standard Proctor energy and modified Proctor energy (independent of placement conditions and typical methods of determining the Cv) and the magnitude of Cv values were compared with the effect of energy ratio of Cv (Cv_{ER}) with reference to the standard energy ratio (SER).

2. Literature review on consolidation studies

McRae [1] developed an index for compaction effort, which is used for the amount of effort required for the compaction of different types of soils. Several researchers like Hilf [2], Jumikis [3], Ring et al. [4], and Wang and Huang [5] have described methods to estimate the optimum water content and maximum dry unit weight of fine-grained soils for standard Proctor compaction test. Benson and Trust [6] conducted the hydraulic conductivity test on thirteen compacted clays compacted to different compaction efforts. Blotz et al. [7] described an empirical method for estimating maximum dry unit weight (γ_d max) and optimum moisture content (OMC) of clayey soils for different compaction energies.

The coefficient of consolidation (Cv) is a vital consolidation characteristic of a soil required during the time rate of consolidation analysis. Terzaghi's one-dimensional consolidation theory [8], Biot's theory of three-dimensional consolidation [9], and large strain consolidation theory (Mikasa [10], Gibson et al., [11], Monte and Krizek [12]) are some of the theories developed to model the complex process of consolidation. However, for minimal strain conditions Terzaghi's one-dimensional consolidation theory can be used for the time rate of consolidation analysis. The study of clay mineralogy on Coefficient of Consolidation by Robinson and Allam [13] showed that for different clay mineralogy of soils, the coefficient of consolidation varies for different pressure ranges and is analyzed through original compression behavior. The increase in Cv values with a variation of pressure for different clay minerals like kaolinite and illite powdered quartz is observed, and the mechanical properties characterize their compressibility behavior. Montmorillonite clay mineral with water as pore fluid within which the compressibility behavior characterized physico-chemical properties of soil, and the variation of Cv with consolidation pressure depends upon compressibility behavior in terms governed by the mechanical or physico-chemical properties of soil. Shiva Prashanth Kumar et al. [14] studied the Coefficient of Consolidation (Cv) in CH soils where Cv for applied pressure was compared with three different methods of determining Cv. The methods involved are Casagrande logarithm of time fitting [15], Taylor's square root of time fitting [16], and inflection point [17]. For all the selected soils, the Cv with applied pressure was correlated by a power law and had the same trend of variation for the pressure range of 50 to 1600 kPa. Madhav and Kurma Rao [18] studied the consolidation characteristics of Kaolinitic clay and concluded that the recovery ratio of a dispersed system is higher than that of a flocculated system irrespective of the pressure increment. Limited experimental data are available in the geotechnical engineering literature illustrating the variation of Cv with consolidation stress. Terzaghi and Peck [19] observed relatively constant Cv values over a wide range of consolidation stress. Leonards and Ramiah [20] observed an upward trend of Cv values for remolded residual clay up to a particular value of consolidation stress and decreased with the consolidation stress exceeding that value. They also noted that the values of Cv for the remolded glacial silty clay continued to increase with consolidation stress.

Table 1 exhibits trends in the variation of coefficient of consolidation with pressure for different clay minerals.

The data suggest that the Cv value decreases for montmorillonite soils and increases for kaolinitic soils with increase in consolidation stress. Robinson and Allam [13] showed that the variation of coefficient of consolidation with the increase in consolidation stress on soils undergoing virgin compression is characterized by mechanical or physico-chemical properties depending upon the dominant clay mineral composing the soil. The literature review on the coefficient of consolidation of soils indicates that limited study has been reported on the coefficient of consolidation of compacted fine-grained soils having different clay mineralogical compositions having different energies imparted.

3. Experiments conducted

Nearly twenty-five soil samples were selected for the experimental program from different locations in Karnataka state, India. These soils were subjected to preliminary laboratory investigation for index properties soils involving Atterberg limits, specific gravity [24], free swell index [25], and grain size analysis [26]. The Atterberg limits were determined using the Casagrande percussion method [27], shrinkage limit [28], and the nature of their clay mineralogical composition was judged by the free swell ratio technique (Prakash and Sridharan [29]). Six field soils were finalized out of 25 soils based on the liquid limit and clay mineralogy, and one commercially available clay mineral was chosen i.e. China clay for the requirement of pure kaolinite clay

Soil type	Dominant clay mineral	Liquid limit: %	Plasticity index: %	Variation of Cv with σ'	Reference
Bentonite	Montmorillonite	118.0	72.0	Decrease	Samarasinghe et al., [21]
Kaolinite	Kaolinite	—	—	Increase	
Kaolinite	Kaolinite	49.0	11.8	Increase	Sridharan et al. [22]
Coarse kaolinite	Kaolinite	48.0	12.4	Increase	Sridharan and Prakash [23]
Fine kaolinite	Kaolinite	46.8	17.4	Increase	
Black cotton soil-2	Montmorillonite	100.8	48.9	Decrease	
Bentonite	Montmorillonite	393.4	343.3	Decrease	
Kaolinite	Kaolinite	53.0	21.0	Increase	Robinson and Allam [13]
Montmorillonite	Montmorillonite	321.0	263.0	Decrease	

Table 1.
Variation of cv with σ' and clay mineral type.

mineral. The selected soils were classified into 3 different groups (G-1, G-2, and G-3) based on the ascending order of liquid limit.

Soils of liquid limit 46% (G-1) (W_L < 50%)

1. Field soil from Bogadi (passing 425 μm sieve), Mysuru District, which contains kaolinite as the predominant clay mineral.

2. Field soil from Nanjangud (passing 425 μm sieve), Mysuru District, which is a montmorillonitic soil.

- Soils of liquid limit 55% (G-2) (50% < W_L < 60%)

3. Field soil from Kollegala, (passing 425 μm sieve), Chamarajanagar district, which contains kaolinite as the predominant clay mineral.

4. Field soil from Kuderu, (passing 425 μm sieve), Chamarajanagar district, which is a montmorillonitic soil.

- Soils of liquid limit 68% (G-3) (60% < W_L < 70%)

5. Field soil from Bannur, (passing 75 μm sieve), Mysuru District, in which both kaolinite and montmorillonite clay minerals are dominant.

6. Field soil from CFTRI lay out, (passing 75 μm sieve), Mysuru District, which contains montmorillonite as the predominant clay mineral.

7. Commercially available clay mineral China clay (representing the kaolinitic soil passing 75 μm sieve) obtained from Seema Chemicals, Bangalore.

Table 2 shows the physical and index properties of the soil types discussed above.

3.1 Compaction tests

Standard and modified compaction tests were conducted for the soils under study (Standard or Light Compaction [30]), and (Modified or Heavy Compaction [31]). For the compaction tests, around 6 to 9 samples of 2.5 kg were taken and mixed with different percentages of moisture contents kept for a gestation period of 5 to 7 days. After placing the soil samples for saturation period, the Proctor compaction tests were conducted on the soil samples to achieve maximum dry density and optimum moisture content.

3.2 Consolidation tests on compacted soils

3.2.1 Sample preparation for consolidation testing

The consolidation ring has a 6 cm diameter (internal) and 2 cm depth or height, and inside the ring, the silicon grease was applied before it is going to compact with soil, which reduces the friction between the soil and ring when the pressure is applied.

S. No.	Soil	Specific Gravity	Liqui d limit (w_L): (%)	Plastic limit (w_P): (%)	Plasticity index (I_p): (%)	Shrinkage limit (w_S): (%)	Grain Size Distribution			IS classification	Clay Mineralogy
							Clay size (%)	Silt size (%)	Sand size (%)		
1	G-1	2.61	46	22	24	13.7	13.0	16	71	CI	Kaolinitic (K- soil)
2		2.70	46	23	23	18.7	7.5	19.5	73	CI	Montmorillonitic (M-soil)
3	G-2	2.74	55	26	29	15.9	37.0	34.5	28.5	CH	Kaolinitic (K- soil)
4		2.85	55	26	28	11.5	39.0	21.0	40.0	CH	Montmorillonitic (M-soil)
5	G-3	2.69	68	30	38	16.1	45.0	55.0	—	CH	Kaolinitic-Montmorillonitic (K-M soil)
6		2.72	68	33	35	13.9	51.0	49.0	—	MH	Montmorillonitic (M-soil)
7		2.67	68	30	38	24.8	63.0	37.0	—	CH	Kaolinitic (K- soil)

Table 2.
Physical and index properties of soils studied.

The consolidation tests were done at three levels of initial molding water contents—corresponding to γ_d max (i.e., OMC), 0.95 γ_d max on the dry side of optimum and 0.95 γd max on the wet side of optimum. The soil sample in the consolidation ring is compacted with the required moisture content and maximum dry density in the consolidation ring and then assembled with a consolidation cell to be positioned with consolidation equipment. The consolidation ring with compacted soil sample with required molding water content and dry density was assembled in its position on the consolidation cell. The cell used for the laboratory experimentation is a permanent or fixed ring type with drainage paths on two sides of the cell, and it has the facility to conduct the falling head permeability test on the soil sample. The consolidation ring has a 6 cm diameter (internal) and 2 cm depth or height, and inside the ring, the silicon grease was applied before it is going to compact with soil, which reduces the friction between the soil and ring when the pressure is applied.

3.2.2 Load: deformation - time measurements for compacted soils

Consolidation tests were conducted according to [32]. The soil samples are kept in desiccator saturation until they reach the optimum moisture content. M-soil sample exhibited swelling on the addition of water into the consolidation cell. In such cases, time-swelling readings were recorded till the equilibrium was reached. The loads applied on soil samples vary from 0.0625 kg/cm^2 (6.25 kPa) (after the permeability measurements were taken) to 16 kg/cm^2 (1600 kPa) with an increment ratio of 1. Time-compression readings were recorded under each consolidation stress increment until the near-equilibrium state was reached. The samples were unloaded upon arriving at the ultimate loading i.e., 1600 kPa in stages (1/4th). Then the samples were dismantled and weighed and their final heights were measured.

The following five typical methods were chosen for computing coefficient of consolidation (Cv) because of their in-built merits:

1. Casagrande Method [15]

2. Taylor's Method [16]

3. Log-log Method [23]

4. Rectangular Hyperbola Method [33]

5. One-point Method [34]

3.2.3 Casagrande method

Time of settlement at 50% degree of consolidation [T_{50} = 0.197] is considered for the proposed method. The test was carried out up to the end of primary consolidation. The variation of time-compression data with typical S-curve shapes is more useful than the other variations. The t_{50} corresponding to the 50% degree of consolidation was calculated using Δ_{50} & Δ_i (Initial compression), and it is estimated using the equation below

$$\Delta_{50} = \left[\frac{\Delta_{100} - \Delta_i}{2} + \Delta_i\right] \quad (1)$$

$$\text{Cv} = 0.197\frac{H^2}{t_{50}} \quad (2)$$

3.2.4 Taylor's method

Time of settlement: at 50% degree of consolidation [T_{90} = 0.848] compression can be found at 0 and 100% primary consolidation. Ninety percent consolidation time varies with different specimens and with different thickness of specimens, but the compression dial is read frequently for a period of up to 1 hr. after initiation of loading. For no initial compression and negative compression this method is not useful. Here 100% primary consolidation is 1/4th of more than the difference in dial gauges compression readings corresponding to 0 and 90% consolidation. For rapid consolidation soils, it is not accurate. The cv value by Taylor's method yields more than the Casagrande method because it is considering T_{50}. Values of Casagrande to Taylor are varying from 0.2 to 1. Leonard [20] reported that the effect of secondary compression may strongly influence the value at t_{90} obtained from $\sqrt{t}$, so Δ_{100} obtained from Casagrande is more reliable.

$$\text{Cv} = 0.848\frac{H^2}{t_{50}} \quad (3)$$

For initial settlement => Taylor's method is reliable.
For Primary consolidation (Δ_{100}) => Casagrande is reliable.

3.2.5 Log-Log method

During some particular stages of the consolidation process, different soils may respond in a way that is consistent with Terzaghi's theory (percentage of consolidation). This might be one of the reasons for the complications that the existing curve-fitting techniques experience when trying to calculate the coefficient of consolidation, Cv. When compared to other techniques, it has been demonstrated that the slope of the early linear component of the theoretical log U-log T curve is constant throughout a greater range of degrees of consolidation, U. In the proposed method, the initial portion of the graphical construction log U-log T i.e. straight line portion is well defined and the plot arrived at log U = 100% degree of consolidation at T = π/4, which is corresponding to U = 88.3%.

$$\text{Cv} = \frac{\pi}{4}\frac{H^2}{t_{88.3}} \quad (4)$$

The method is more flexible; it is evident when identifying the characteristic straight lines and more accurate when these lines were intersecting each other. It does not rely on initial compression to calculate C_V.

3.2.6 Rectangular hyperbola method

The variation of t/Δ vs. time was plotted (Measuring values of Slope (m) and intercept (C)) to determine the Cv at 60% degree of consolidation.

$$Cv = 0.24\frac{mH^2}{c} \tag{5}$$

From R-H Method Δ_{100} can also be obtained for 100% primary consolidation.

$$\Delta_{100} = H_0\left[0.862\frac{1}{mH_0} - 3.677 * 10^{-4}\right] \tag{6}$$

Soils with secondary compression show two straight lines and the initial compression cannot be estimated from this method directly.

3.2.7 One point method

In response to the finding that the experimental behavior of soil without consideration of initial and secondary compression effects fits closely to the theory in the range of 40% < U < 60%. The value of the final compression Δ_{100} is taken into account at the end of the loading period in order to calculate Cv. Each loading increment is often taken to be 24 hours long. To obtain the value of final compression, one must therefore wait 24 hours or longer and the compression equal to 50% consolidation (Δ_{50}), or 0.5 Δ_{100}, and the associated time t_{50}, is calculated. According to the mentioned time t_{50}, Cv is calculated.

$$\Delta_{50} = \frac{\Delta_{100}}{2} \tag{7}$$

$$Cv = 0.197\frac{H^2}{t_{50}} \tag{8}$$

4. Results and discussions

4.1 Coefficient of consolidation (Cv) of compacted fine-grained soils with variation of pressure based on liquid limit

The Coefficient of Consolidation (Cv) is a parameter that is being generally considered in estimating the Settlement characteristics. The Coefficient of consolidation (Cv) is determined from typical methods chosen for the present experimental study have been categorized into three different clay mineralogical soil groups of kaolinite, montmorillonite and kaolinite-montmorillonite having different liquid limit range (46, 55, and 68%) with the pressure varying from 6.25 to 1600 kPa.

Table 3 shows that the range of Cv values for the kaolinite, montmorillonite, and kaolinite-montmorillonite soils under study having the liquid limit range of 46, 55, and 68% and for the pressure ranging from 6.25 to 1600 kPa.

Coefficient of consolidation (Cv) (cm^2/s)		
K-soils	**M-soils**	**K-M soils**
1.28×10^{-7} to 8.21×10^{-1}	1.86×10^{-7} to 9.76×10^{-1}	1.32×10^{-7} to 8.7×10^{-1}

Table 3.
Coefficient of consolidation values of K-soils, M-soils, and K-M soils.

4.2 Variation of coefficient of consolidation (Cv) with respect to energy levels

The role of compaction energy on the engineering behavior of fine-grained soils is well documented in geotechnical literature. However, no attempt has been made to correlate the consolidation characteristics like Cv to compaction energy levels of different clay mineralogical fine-grained soils. In the documented literature, detailed discussions were made concerning desired changes in the engineering properties of soil due to the effect of compaction energy imparted on the soil. Further, the heavy compaction energy level to light compaction energy level is 4.54 (Standard Energy ratio (SER)). The variation of Cv (Light compaction) with Cv (Heavy compaction) has not been made in the past, as seen from the documented literature with particular reference to the clay mineralogy, placement condition, and pressure concerned.

Amid the consolidation characteristics illustrated in the documented geotechnical literature, the Cv is an extremely variable parameter and the number of methods (27) are highlighting the same. The computation of Cv values for the soil samples for different energy levels and clay mineralogy consumes lot of time and associated cost. The present experimental approach has been proposed to estimate the Cv values for heavy compaction energy level through light compaction energy level is of paramount importance from an economic perspective, which is illustrated in **Figures 1–10**.

Figures 1–10: The values have been taken into consideration in the logarithmic (base 10) scale to provide a clear depiction of the cv values on the ordinate and abscissa. In view of the enormous amount of experimental work and data, three reference pressures—50, 100, and 200 kPa were selected for soils with liquid limits of 46, 55, and 68%, respectively. This is because most sub-structures, especially those for

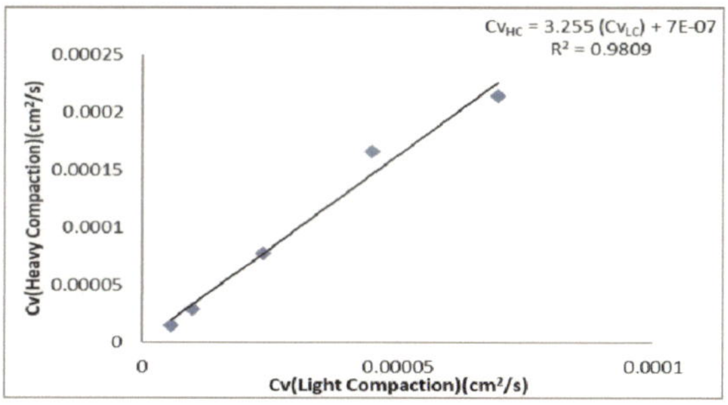

Figure 1.
Correlation between cv (light compaction) and cv (heavy compaction) of K-soil (W_L = 46%) (50 kPa).

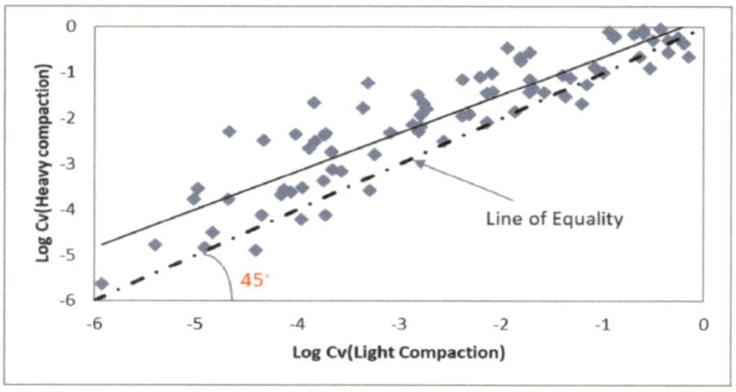

Figure 2.
Correlation between cv (light compaction) and cv (heavy compaction) of K-soil (W_L = 46%).

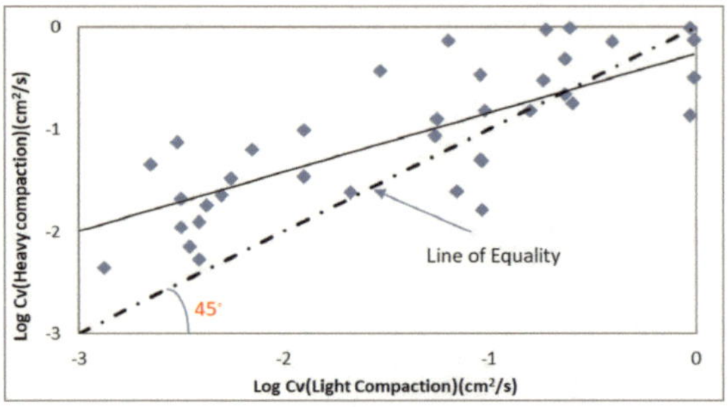

Figure 3.
Correlation between cv (light compaction) and cv (heavy compaction) of M-soil (W_L = 46%).

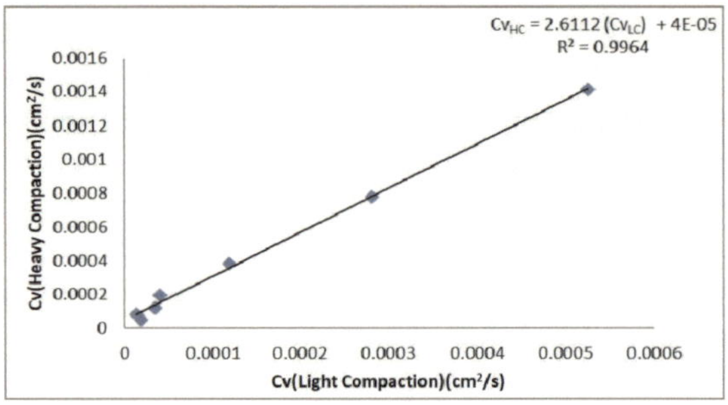

Figure 4.
Correlation between cv (light compaction) and cv (heavy compaction) of M-soil (W_L = 55%) (100 kPa).

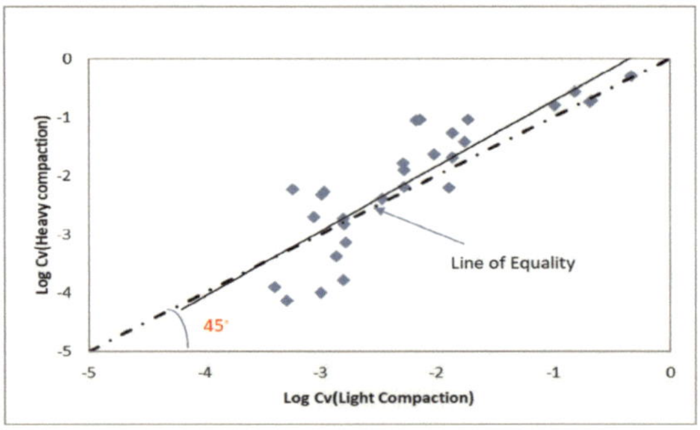

Figure 5.
Correlation between cv (light compaction) and cv (heavy compaction) of K-soil (W_L = 55%).

shallow foundations, are required to have a minimum safe bearing capacity of 100 kPa, which is typically determined by shear and settlement criteria, with allowable settlements of ≤25 and ≤ 40 mm, respectively. From this, a comprehensive understanding of the variation of Cv with pressure with respect to the light compaction and heavy compaction energy levels has been achieved.

Figure 1 shows the variation of Cv (Light compaction) with Cv (Heavy compaction) for the soil having liquid limit of 46% (kaolinitic soil) from the chosen methods of determining Cv (@ 50 kPa pressure).

Figures 2 and **3** demonstrate the correlation between Cv (Light compaction) and Cv (Heavy compaction) values of combined typical methods of determining Cv chosen for the study and placement conditions of soils having a liquid limit of 46%

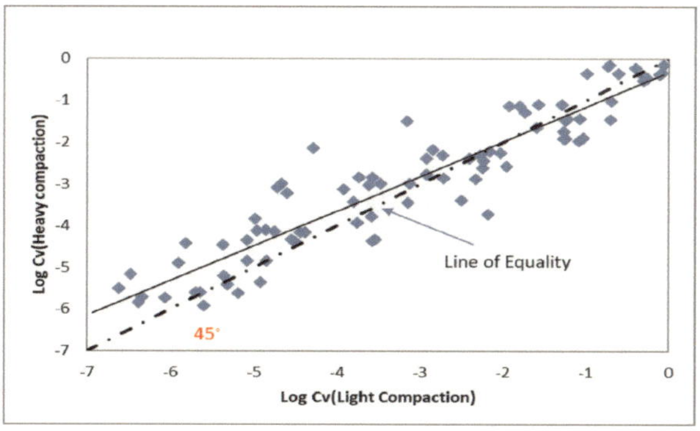

Figure 6.
Correlation between cv (light compaction) and cv (heavy compaction) of M-soil (W_L = 55%).

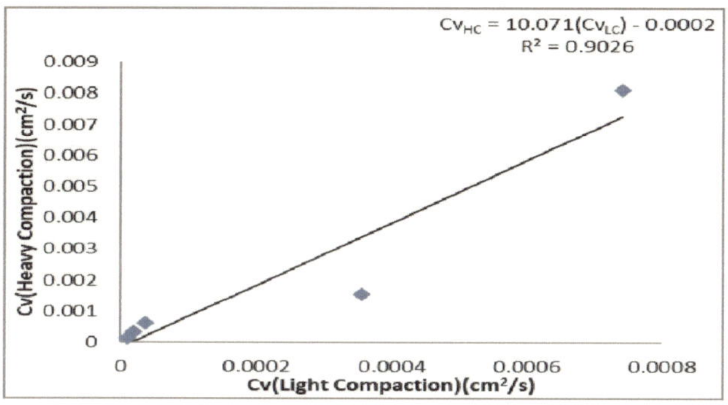

Figure 7.
Correlation between cv (light compaction) and cv (heavy compaction) of K-M soil (W_L = 68%) (200 kPa).

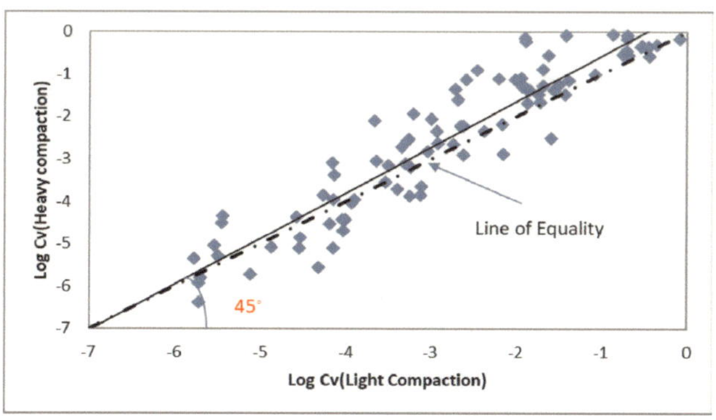

Figure 8.
Correlation between cv (light compaction) and cv (heavy compaction) of K-soil (W_L = 68%).

(kaolinite and montmorillonite soil). The values of Cv have been compared with the line of equality to observe the deviation of values from the correlation equation.

Eq. (9) & Eq. (10) represents that, the relationship between values of Cv for heavy compaction to the light compaction energy level for the soils having different clay mineralogy and the liquid limit of 46%.

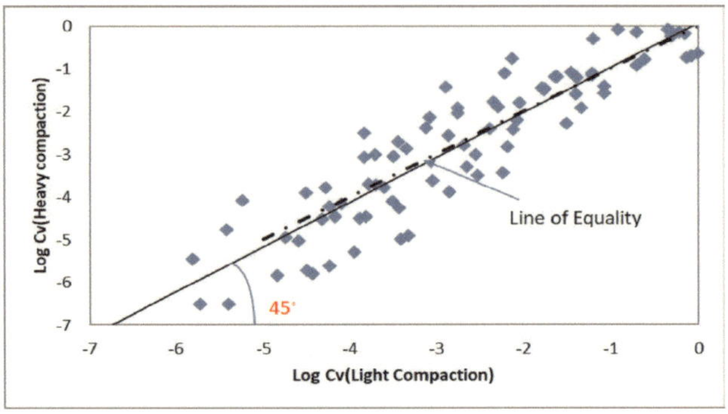

Figure 9.
Correlation between cv (light compaction) and cv (heavy compaction) of M soil (W_L = 68%).

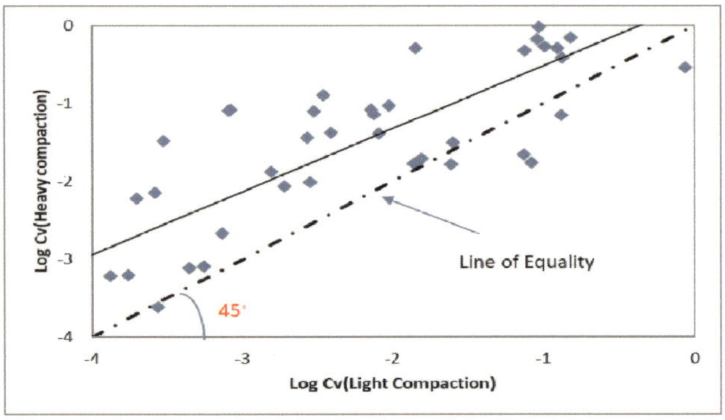

Figure 10.
Correlation between cv (light compaction) and cv (heavy compaction) of K-M soil (W_L = 68%).

$$Cv(Heavy) = 0.8371Cv(Light) + 0.1825 \ R^2 = 0.90 \text{ For K} - \text{Soil} \quad (9)$$

$$Cv(Heavy) = 0.5755Cv(Light) - 0.2672 \ R^2 = 0.91 \text{ For M} - \text{soil} \quad (10)$$

Figure 4 shows the variation of Cv (Light compaction) with Cv (Heavy compaction) for the soil having liquid limit of 55% (montmorillonitic soil) from the chosen methods of determining Cv (@ 100 kPa pressure).

Figures 5 and **6** demonstrate the correlation between Cv (Light compaction) and Cv (Heavy compaction) values of combined typical methods of determining Cv chosen for the study and placement conditions of soils having a liquid limit of 55% (Kaolinite and Montmorillonite soil). The values of Cv have been compared with the line of equality to observe the deviation of values from the correlation equation.

Eq. (11) & Eq. (12) represents that the relationship between values of Cv for heavy compaction to the light compaction energy level for the soils having different clay mineralogy of same liquid limit (55%).

$$Cv(Heavy) = 1.1138Cv(Light) + 0.3846 \ R^2 = 0.86 \text{ For K} - \text{Soil} \quad (11)$$

$$Cv(Heavy) = 0.8337Cv(Light) - 0.3101 \ R^2 = 0.93 \text{ For M} - \text{Soil} \quad (12)$$

Figure 7 shows the variation of Cv (Light compaction) with Cv (Heavy compaction) for the soil having liquid limit of 68% (kaolinitic-montmorillonitic soil) from the chosen methods of determining Cv (@ 200 kPa pressure).

Figures 8–10 demonstrate the correlation between Cv (Light compaction) and Cv (Heavy compaction) values of combined typical methods of determining Cv chosen for the study and placement conditions of soils having a liquid limit of 68% (kaolinite, montmorillonite, kaolinite-montmorillonite soil). The values of Cv have been compared with the line of equality to observe the deviation of values from the correlation equation.

Eq. (13), Eq. (14), and Eq. (15) show the relationship between values of Cv for heavy compaction to the light compaction energy level for the soils having different clay mineralogy and the liquid limit of 68%.

$$Cv(Heavy) = 1.0783Cv(Light) + 0.5126\ \mathrm{R}^2 = 0.93\ \mathrm{For\ K - Soil} \tag{13}$$

$$Cv(Heavy) = 1.051Cv(Light) + 0.0864\ \mathrm{R}^2 = 0.91\ \mathrm{For\ M - Soil} \tag{14}$$

$$Cv(Heavy) = 0.8077Cv(Light) + 0.288\ \mathrm{R}^2 = 0.92\ \mathrm{For\ K - M\ Soil} \tag{15}$$

From **Figures 1–10** and Eq. (1) through Eq. (7), it can be observed that the values of Cv (heavy compaction) can be estimated by substituting the values of Cv (light compaction) for combined placement conditions of soils having different clay mineralogy with a fair degree of accuracy of regression values ranging from 0.86 to 0.99, irrespective of the method of determining Cv and placement condition. Furthermore, the values of Cv from the equation are closer to the line of equality in all of the soils that were investigated, with a lesser degree of deviation.

Based on the correlation equations derived from the present experimental study, the values of Cv can be directly estimated for the Reduced Standard Proctor (RSP) and Reduced Modified Proctor (RMP) energy level also, by considering the amount of energy to 60 percent of actual energy in the Modified and Standard Proctor effort, respectively (RSP = 60% of Standard Proctor Energy & RMP = 60% of Modified Proctor Energy).

4.3 Variation of energy ratio with coefficient of consolidation

The standard energy ratio (SER) derived from Proctor lubrication theory [35] i.e., the value of compaction energy determined from the heavy compaction energy to the value of compaction energy determined from the light compaction energy. Here the energy has been considered directly to determine the energy ratio. The defined SER equation represented in Eq. (16)

$$\text{Standard energy ratio (SER)} = \frac{\text{Heavy compaction energy [IS : 2720 − Part 8 (1983)][2703 KJ}/m^3]}{\text{Light compaction energy[IS : 2720 − Part 7(1980)][596 KJ}/m^3]} \tag{16}$$

From the reference of the SER, the new equation has been defined to compare in terms of coefficient of consolidation (Cv) point of view i.e., Energy ratio of Cv (CvER).

The Cv_{ER} represents the variation of Cv values with respect to the effect of energy ratio and pressure. Here, the values of Cv considered for the estimation of the energy ratio of Cv (Cv_{ER}) are the combined values of typical methods of determining Cv mentioned in the methodology and the placement conditions of dry side of optimum, optimum, and wet of optimum, for the individual pressures, respectively.

The energy ratio of Cv and average ratio are defined below in Eq. (17) & Eq. (18)

$$\text{Energy ratio of Cv (CvER)} = \frac{\text{Cv Value of Heavy compaction energy level}}{\text{Cv value of Light compaction energy level}} \tag{17}$$

$$\text{Average ratio } (Cv_{AR}) = \text{The average of values of } Cv_{ER} \tag{18}$$

The Cv_{ER} represents the variation of Cv values with respect to the effect of energy ratio and pressure as shown below in **Table 4** for M-Soil having W_L = 46% (100 kPa).

Similar observations were made for the soils having different pressure range, clay mineralogy, and liquid limit.

Figures 11–13 show the relation between the average ratio of Cv values with respect to the pressures ranging from 50 to 1600 kPa for the soils having different clay mineralogy of liquid limits 46, 55, and 68%, respectively and have been compared with standard energy ratio (SER) i.e. 4.54.

Tables 5 and **6** show the relation between the average ratio of Cv values with respect to the pressures ranging from 50 to 1600 kPa for the soils having different clay mineralogy of liquid limits 46, 55, and 68%, respectively.

Table 5 shows the variation of Cv_{AR} with the consolidation stress.

Table 6 shows the variation of the percentage amount of energy achieved (defined from SER) at gradual consolidation loading ranging from 50 to 1600 kPa, for the soils under study. (Eq. (19))

Cv (Light Compaction)	Cv (Heavy Compaction)	Cv_{ER}	*Cv_{AR}
1.07E-04	0.000661	6.16595	8.22
1.15E-04	0.001318	11.48154	
4.37E-04	0.00195	4.466836	
9.55E-04	0.002951	3.090295	
0.000295	0.005495	18.62087	
0.001	0.005495	5.495409	

Table 4.
Estimation of the average ratio of cv values for the M-soil at 100 kPa pressure (WL = 46%).

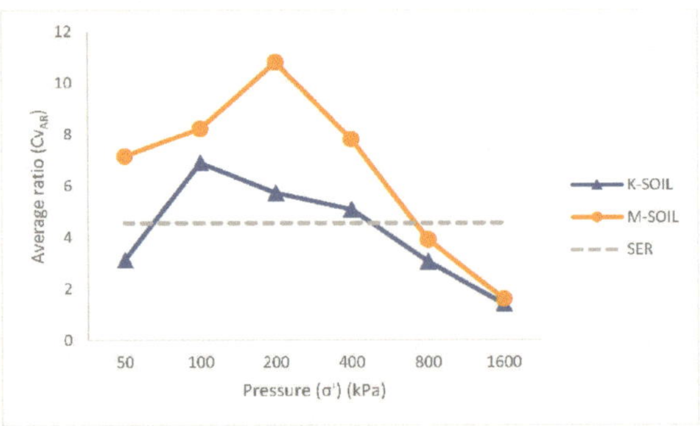

Figure 11.
Variation of Cv_{AR} with pressure (W_L = 46%).

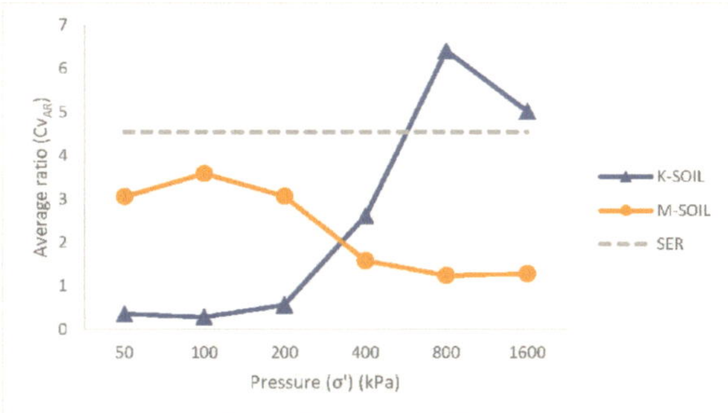

Figure 12.
Variation of Cv_{AR} with pressure (W_L = 55%).

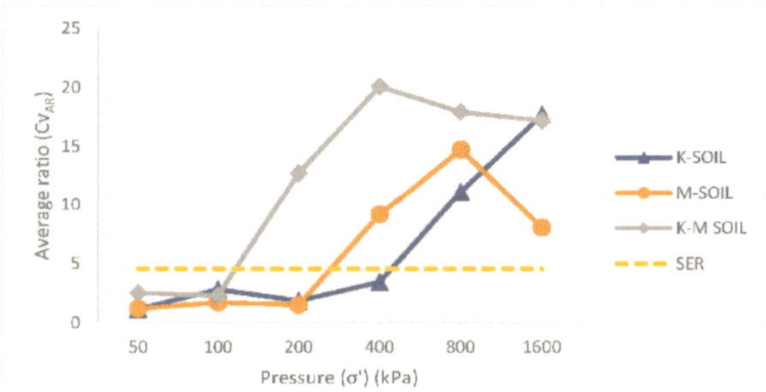

Figure 13.
Variation of Cv_{AR} with pressure (W_L = 68%).

$$\text{Percentage of energy achieved } (\%) = \frac{\text{Cv}_{\text{AR}}}{\text{SER}} * 100 \tag{19}$$

From **Tables 5** and **6**, it can be observed that as the liquid limit of soils increases the average ratio of Cv values increases for high liquid limit soils (W_L = 68%) (K, M, & K-M soils) having higher average ratio values (1.13 to 17.66, 1.17 to 14.7, and 2.28 to 20.05) and percentage amount of energy achieved (24.89 to 388.98, 25.77 to 176.8 and

Pressure (kPa)	**Cv_{AR}**							
	W_L = 46%		**W_L = 55%**		**W_L = 68%**			
	K-soil	**M-soil**	**K-soil**	**M-soil**	**K-soil**	**M-soil**	**K-M soil**	**SER**
50	3.12	7.14	0.35	3.05	1.13	1.17	2.51	4.54
100	6.88	8.22	0.28	3.58	2.8	1.69	2.28	4.54
200	5.72	10.8	0.56	3.06	1.84	1.49	12.71	4.54
400	5.08	7.8	2.6	1.57	3.42	9.16	20.05	4.54
800	3.05	3.9	6.4	1.23	11.1	14.7	17.89	4.54
1600	1.4	1.6	5.01	1.27	17.66	8.03	17.16	4.54

Table 5.
Values of average ratio Cv_{AR}.

Pressure	Percentage amount of energy achieved @ Gradual consolidation loading						
	G-1 (W_L = 46%)		G-2 (W_L = 55%)		G-3 (W_L = 68%)		
	K-soil	M-soil	K-soil	M-soil	K-soil	M-soil	K-M soil
50	68.72	157.27	7.71	67.18	24.88	25.77	55.28
100	151.54	181.05	6.17	78.85	61.67	37.22	50.22
200	125.99	237.88	12.33	67.40	40.52	32.82	279.96
400	111.89	171.80	57.27	34.58	75.33	201.76	441.63
800	67.18	85.90	140.97	27.09	244.49	323.78	394.05
1600	30.84	35.24	110.35	27.97	388.98	176.87	377.97

Table 6.
Variation of the percentage of energy achieved with reference to SER for the gradual consolidation loading or stress.

55.28 to 377.97), respectively, than other soils pertaining to liquid limits 46 and 55% respectively for the same pressure range. Here the maximum value of Cv_{AR} is referred to as the maximum percentage of the amount of energy achieved at the gradual consolidation loading and it is considered as the percentage amount of energy achieved in terms of standard energy ratio (the soils compacted at dynamic loading for different energy levels) for the individual consolidation stress.

For the soil having W_L = 46%, the values of Cv_{AR} lies above the SER line for the pressure ranging from 100 kPa to 400 kPa (K-soil) and it is ranging from 50 kPa to 400 kPa for the M-soil, respectively (**Figure 11**). For the soil having W_L = 55%, the values of Cv_{AR} lies above the SER line for the pressure ranging from 800 kPa to 1600 kPa (K-soil) and it lies below the SER line, irrespective of the pressure range for the M-soil, respectively (**Figure 12**). For the soil having W_L = 68%, the values of Cv_{AR} are lying above the SER line for the pressure ranging from 200 to 1600 kPa (K-M Soil), 400 to 1600 kPa (M-soil), and 800 to 1600 kPa (K-soil), respectively (**Figure 13**).

From **Figures 11–13** and **Tables 5** and **6**, it can be observed that the Cv_{AR} is having a decrease in tendency beyond 200 kPa (M-soil) and 400 kPa (K-soil), respectively, and reaches its minimal value at 1600 kPa for the W_L = 46%, the Cv_{AR} is having a decrease in tendency beyond 100 kPa (M-soil) and 800 kPa (K-soil) for the W_L = 55% and the W_L = 68%, it is beyond 800 kPa for M-soil and 400 kPa for K-soil i.e., the decrease in the trend of variation of Cv_{AR} is changing with the change in the liquid limit, and the percentage amount of energy achieved with respect to the pressure is also increasing with the increase in the liquid limit. The behavior of Cv_{AR} for K-soils is in decreasing trend for the 46% liquid limit, whereas for the 55 and 68% liquid limits, the increase in trend was observed. The occurrence is ascribed to the increase in the amount of percentage of fines when the liquid limit of the soil increases. For the soil having liquid limit of 68%, the void ratio increases with an increase in the percentage of fines i.e., for the soils having liquid limit 46% and 55% are having less percentage of fines having less void ratio compared to the 68% liquid limit soil irrespective of the clay mineralogy and the coefficient of consolidation achieved with the increase in pressure at the heavy compaction energy level is higher due to the higher compression of voids affects the increase in the values of Cv at heavy compaction energy level which in turn leading to the Cv_{AR}.

From **Figures 11–13** and **Tables 5** and **6**, it can be also observed that, for W_L = 46%, the maximum value of Cv_{AR} for M-soil is having greater magnitude than the Cv_{AR} for K-soil. The maximum value of Cv_{AR} for M-soil is observed at a pressure of 200 kPa whereas for K-soil it is 100 kPa. For W_L = 55%, the maximum value of Cv_{AR} for M-soil is having greater magnitude than the Cv_{AR} for K-soil. The maximum value of Cv_{AR} for M-soil is observed at a pressure of 100 kPa whereas for K-soil it is 800 kPa. For W_L = 68%, The maximum value of Cv_{AR} for K-M soil is having greater magnitude than the Cv_{AR} for K and M-soil. The maximum values of Cv_{AR} were observed at 1600, 800, and 400 kPa, for K, M, and K-M soils respectively.

The mechanism of the energy ratio concept can be attributed to the fact that, according to Proctor's capillarity and lubrication theory [35], water has a dual effect of capillarity (or suction) and lubrication. It is observed that due to high capillarity, the dry density is lower for dry soil and as water is added, the capillarity is reduced, and water also lubricates the particle interaction, giving rise to increased dry density up to the maximum dry density. It is depending upon the amount of energy imparted on the soil (Heavy and Light compaction) and also the pressure creates additional effort on the consolidation specimen with respect to a decrease in void ratio due to an increase in the rate of expulsion of water from soil voids. This behavior leads to an increase in the rate of compression and it is defined in the form of a coefficient of consolidation with the effect of energy ratio.

5. Conclusions

Based on a detailed experimental study of compacted fine-grained soils having different clay mineralogy, placement conditions, and energy levels, the following conclusions can be made:

- The magnitudes of Cv values computed by different user-friendly methods for K, M, and K-M soils vary from 1.28×10^{-7} to 9.76×10^{-1} cm^2/s, and M-soils exhibit a higher magnitude of Cv in relative comparison to K-soils and K-M soils by virtue of clay mineralogy.

- The Cv values of heavy compaction energy level can be estimated directly by knowing the Cv values of light compaction energy level only through correlation equations irrespective of the method of determining Cv and placement conditions (it is independent of the method of determining Cv of five user-friendly methods used in the experimental study).

- The Cv_{AR} values increase with an increase in pressure up to 400 kPa for W_L = 68% (K, M & K-M soils).

- The Cv_{AR} has a decreasing tendency beyond the pressure range of 400 kPa for K-soils and 200 kPa for M-soils (W_L = 46%), 800 and 100 kPa for K and M soils (W_L = 55%) and it is 400 and 800 kPa for K-M and M-soil, respectively.

- The maximum value of (Cv_{AR}) M-SOIL > Maximum value of (Cv_{AR}) K-SOIL (W_L = 46 & 55%), Maximum value of (Cv_{AR}) K-M SOIL > Maximum value of (Cv_{AR}) M-SOIL > Maximum value of (Cv_{AR}) K- SOIL (W_L = 68%).

Conflict of interest

The authors declare no conflict of interest.

Author details

Heggadadevanakote Subbarao Prasanna and Unnam Anil*
Department of Civil Engineering, NIE, Mysuru, Karnataka, India

*Address all correspondence to: anil.nanna441@gmail.com

References

[1] McRae JL. Index of compaction characteristics. In: Symposium on Application of Soil Testing in Highway Design and Construction, ASTM, STP. Vol. 239. Philadelphia: ASTM International; 1958. pp. 119-127

[2] Hilf J. A Rapid Method of Construction Control for Embankments of Cohesive Soil, Engrg. Monograph No. 26. USA: U.S. Bureau of Reclamation, Denver, CO; 1956

[3] Jumikis AR. Geology and soils of the Newark (NJ) metropolitan area. Journal of the Soil Mechanics and Foundations Division. 1958;**94**(2):1646-1

[4] Ring GW, Sallberg JR, Collins WH. Correlation of Compaction and Classification Test Data, Bulletin 325. Washington DC, United States: Highway Research Board; 1961. pp. 55-75

[5] Wang M, Huang C. Soil compaction and permeability prediction for models. Journal of Environmental Engineering. 1984;**110**:1063-1083

[6] Benson CH, Trust JM. Hydraulic conductivity of thirteen compacted clays. Clays and Clay Minerals. 1995;**43**: 669-681

[7] Blotz L, Benson C, Boutwell G. Estimating optimum water content and maximum dry unit weight for compacted clays. Journal of Geotechnical and Geo-environmental Engineering. 1998;**124**:907-912

[8] Terzaghi K. Principle of soil mechanics. Engineering News-Record. 1925;**95**:742-746, a series of articles

[9] Biot MA. General theory of three-dimensional consolidation. Journal of Applied Physics. 1941;**12**:155-164

[10] Mikasa M. The Consolidation of Soft Clay- a New Consolidation Theory and its Application. Tokyo Japan: Kajima institution Publishing Co., Ltd; 1963

[11] Gibson RE, England GL, Hussey MJL. The theory of one-dimensional consolidation of saturated clays, finite non-linear consolidation of thin homogeneous layers. Canadian Geotechnical Journal. 1967;**17**(3):261-227

[12] Monte JL, Krizek RJ. One dimensional mathematical model for large-strain consolidation. Géotechnique. 1976;**26**:495-510

[13] Robinson RG, Allam MM. Analysis of consolidation data by a non-graphical matching method. Geotechnical Testing Journal. 1998;**21**(2):140-143

[14] Kumar SP. Evaluation of coefficient of consolidation in CH soil. Jordan Journal of Civil Engineering. 2016;**10**(4): 515-528

[15] Casagrande A, Fadum RE. Notes on Soz7 Testing for Engineering Purposes, Soil Mechanics Series No. 8. Vol. 268. Cambridge, MA: Harvard University, Graduate School of Engineering; 1940. p. 1939

[16] Taylor DW. Research on Consolidation of Clays, Serial 82. USA: Massachusetts Institute of Technology, Department of Civil and Sanitary Engineering; 1942

[17] Cour FF. Inflection point method for computing cv. Journal of the Soil Mechanics and Foundation Engineering Division, ASCE. 1971;**97**(5):827-831

[18] Madhav MR. Evaluation of the coefficient of consolidation – A

numerical method. The Journal of Institution of Engineers (India). 1964; **44, No. 11**(Pt. CI6):679-684

[19] Terzaghi K, Peck RB. Soil Mechanics in Engineering Practice. 2nd ed. New York: Wiley; 1967

[20] Leonards GA, Ramaiah BK. Time effects in the consolidation of clay. In: Symposium on Time Rates of Loading in Soil Testing, ASTM, STP. Vol. 254. Philadelphia: ASTM International; 1959. pp. 116-130

[21] Samarasinghe AM, Huang YH, Drnevich VP. Permeability and consolidation of normally consolidated soils. Journal of Geotechnical and Geoenvironmental Engineering Division, ASCE. 1982;**108**(GT6):835-850

[22] Sridharan A, Prakash K. Discussion on limitations of conventional analysis of consolidation settlement. ASCE Journal of Geotechnical Engineering. 1995;**121**(6):517

[23] Sridharan A, Prakash K. The log Δ - log t method for the determination of coefficient of consolidation, Geotechnical Engineering (London),. Vol. 125. 1997. Paper 10911, pp. 27-32

[24] IS: 2720-Part 3/Sec-1. Indian Standard Methods of Test for Soils: Determination of Specific Gravity. New Delhi: Bureau of Indian Standards; 1980

[25] IS: 2720-Part 40. Indian Standard Methods of Test for Soils: Determination of Free Swell Index of Soil. New Delhi: Bureau of Indian Standards; 1977

[26] IS: 2720-Part 4. Indian Standard Methods of Test for Soils: Grain Size Analysis. New Delhi: Bureau of Indian Standards; 1985

[27] IS: 2720-Part 5. Indian Standard Methods of Test for Soils: Determination of Liquid Limit and Plastic Limit (second revision). New Delhi: Bureau of Indian Standards; 1985

[28] IS: 2720-Part 6. Indian Standard Methods of Test for Soils: Determination of Shrinkage Factors (first revision). New Delhi: Bureau of Indian Standards; 1972

[29] Prakash K, Sridharan A. Free swell ratio and clay mineralogy of fine-grained soils. Geotechnical Testing Journal, ASTM. 2004;**27**(2):220-225

[30] IS: 2720-Part 7. Indian Standard Methods of Test for Soils: Determination of Water Content-Density Relation Using Light Compaction (Second Revision). New Delhi: Bureau of Indian Standards; 1980

[31] IS: 2720-Part 8. Indian Standard Methods of Test for Soils: Determination of Water Content-Dry Density Relation Using Heavy Compaction (Second Revision). New Delhi: Bureau of Indian Standards; 1983

[32] IS: 2720-Part 15. Indian Standard Methods of Test for Soils: Determination of Consolidation Properties. New Delhi: Bureau of Indian Standards; 1986

[33] Sridharan A, Murthy NS, Prakash K. Rectangular hyperbola method of consolidation analysis. Geotechnique. 1987;**37**(3):355-368

[34] Pandian NS, Sridharan A, Kumar KS. A new method for the determination of coefficient of consolidation. Geotechnical Testing Journal, ASTM. 1992;**15**(1):74-79

[35] Proctor R. Fundamental principles of soil compaction. Engineering News Record. 1933;**111**(9):245-248

Chapter 6

Construction Techniques Related to Clay Soils: A Case Study in Africa

David Mubiru and Sam Bulolo

Abstract

Clay, as one of the soils, must be assessed before being used for particular purposes such as cement production, and the assessment involves detailed site investigations conducted through laboratory or *in situ* equipment; thus, over the years, different systems or methods, such as stabilization, earth building, nanotechnology, and investigations methods, have been improved. However, although improvements have been made, there are still challenges and gaps within the systems and processes in terms of construction, consumption, usage, properties, etc.

Keywords: clay, stabilization, soils, materials, strength

1. Introduction

Clay, being the oldest material and most effective material on earth, has been used in various activities, and due to advancement (sixteenth century/1601 AD), humans have managed to use the soils *in site* investigation, foundation, walls, windows and doors, floor, ceiling, roof, cement production, paper making, landscape features, paver bases, house foundation, chemical filtering, brick manufacture, clinker adsorbent, architecture, construction engineering, clinker, and soil stabilization.

So, this is a deposit or fine material formed from weathering of sedimentary rocks described as having particle size of less than 0.002 mm by ASSHTO, USDA, and MIT, and according to UCS, it is classified as fines (silt and clay) less than 0.075 mm; however, it becomes hard when heated and soft when cooled, but according to [1] records, when soil was first used since the ancient past were not available. Technological advancement has resulted in new construction methods because the twenty-first century requires intelligent and sustainable construction techniques to reduce environmental impact and supplement efficiency. However, tracing the history of construction techniques within Africa was tricky since there were no clear boundaries on when each was first used.

1.1 Site investigation

Every infrastructure is supported by soil; hence, its stability and safety depend on the soil properties, which are supposed to be determined through, for example, desk study, site survey, topographic and hydrographic, geological and hazard mapping,

 IntechOpen

subsurface investigation and dilapidation or precondition survey, etc., and it is the main stage in project designs, thus defined as an assessment and evaluation of soil or bedrock characterization.

Sampling, such as standard split spoon, thin-wall tube, and piston sampler, should be done to be able to obtain samples used for the routine; laboratory tests, such as moisture content, plasticity, shrinkage, particle size distribution, density, specific gravity, particle density, compaction, shear strength, permeability, pore water pressure, suction, compressibility, consolidation, standard penetration, cone penetration, pressure meter, flat dilatometer, vane shear, and plate loading, are the *in situ* tests conducted, but engineers in developing countries have resorted to adopting low-standard methods due to funds and technical knowledge. So for the case of laboratory tests, for example, to determine the moisture content, consecutive tests must be conducted by drying simultaneously to a particular weight, at a temperature ranging from 105 to 110°C not exceeding 50°C until there is no further mass loss, then the other laboratory tests such as liquid limit, shrinkage, particle size distribution, particle density, bulk density will be carried out according to the [2, 3] shear box and triaxial tests [4], and permeability tests [5]. Furthermore, *in situ* soil suction can be measured by tensiometers, pore water pressure by vacuum pressure gauge or pressure transducer, compressibility by an oedometer or triaxial, and consolidation by an oedometer.

FVT and CPT/CPTU are generally the most direct and reliable in soft soils, DMT applies to a wide range of soil, PMT is best suited from medium to very stiff clays, SBPMT is challenging to deploy, SPT is not suited to very soft clay, and PLT is mainly used at shallow depths or excavations [6]. In Uganda, the *in situ* method commonly used method is standard penetration (SPT), then shear box and triaxial tests for laboratory testing [7]; however, there is a delay in identifying appropriate methods appropriate for each country within Africa.

1.2 Building technology

This is a collection of methods and technical processes used in construction or architecture, and in Africa, cement has dominated the construction industry, but the earth is a commonly used material in masonry techniques, for example, sun-dried or kilns-fired Adobe, cob, compressed earth blocks, rammed earth construction, earth shielding, wattle, and daub products. Although these are cheap and easy to construct, they historically lead to weak houses, which are affected by floods and seismic events, so techniques, such as stabilization, were introduced in the second half of the last century to increase the strength and durability of the materials.

1.2.1 Earth building

This is a construction technique performed by mixing reasonably dry inorganic subsoil, non-expansive clay, sand, and aggregate, which requires dampening, mechanically pressing at high pressure, and drying the resulting material.

Cob construction is comprised of materials such as clay soil, coarse, straw, and water, after which the mix is used to create walls without formwork or mechanical ramming, but the mixture is applied so that when one course dries, the construction continues.

Adobe blocks are an unburned mud product, molded in pieces and sun-dried, and their manufacture is a straightforward process, and it was began to be used from 10,000 to 8000 BC [8]. A crater-like mound of earth is made on or near the site of the proposed structure, and after pouring water into it, then it is puddled to a plastic consistency by the workers, or adobero tramping barefooted throughout the mass or, better still, by hoeing to ensure a more thorough mixture. Then 1.5-inch or two-inch thick layers of straw or chopped hay, preferably of short lengths, are spread over the top, and the entire mass is again kneaded to distribute the binder uniformly. To prevent the straw from working to the bottom, it should not be added until the material has been well-puddled. The latter, along with the thoroughness with which the puddling is done, largely depends on the reduction of cracks that occur during sun drying and the material's ultimate strength.

Rammed earth construction, initially in North Africa and the Middle East, was a technique used, and soils were compacted between formwork, then removed after the soil has dried, thus resulting in a structural component [9, 10], then wattle and daub is where a frame is created with secondary and primary timbers, and the main timbers support the secondary so that the mud produced is placed unto the frame made.

Traditionally, wooden molds were initially used to make blocks, but over the years, methods of machine technology to produce blocks have progressed, for example, the CINVA-RAM press machine in Bogota-Colombia (1950), machine in Uganda (1990), and a manual interlocking in Kenya, which produces curved, straight double interlocking, wide format interlocking blocks, after which these blocks are used for construction of water tanks, and wall creation, etc. (**Figures 1–6**) [11].

Figure 1.
Compressed earth blocks.

Figure 2.
Adobe blocks.

Figure 3.
Cob.

Figure 4.
Rammed earth.

Figure 5.
Earth sheltering.

Figure 6.
Wattle and daub.

1.2.2 Nanotechnology

This combines polymers (natural or synthetic), nano clay, and polymers are substances of larger molecules comprised of several chemical units named macromolecules. The nano clay differs by the different minerals, such as montmorillonite, bentonite, kaolinite, hectorite contained within, and there are three synthesis methods, for example, solution blending, melt blending, and *in situ* polymerization. Solution blending yields better results than melt blending due to the proper dispersion of the clay within the polymer matrix, and due to its low viscosity and high agitation power, but melt blending is considered industrially viable and eco-friendly, with high economic potential. Generally, *in situ* polymerization is widely used.

Solution blending is a combination of a polymer or prepolymer, clay in a solvent such as water, chloroform, or toluene, after which the polymer chains intercalate and displace the solvent within the interlayer of the clay, thus polymer/nano clay composites will be formed. The process is comprised of three stages, namely the dispersion of clay in a polymer solution, controlled removal of the solvent, and composite film casting.

Melt blending is where the required amount of intercalated nano clay particles are mixed with polymers at temperatures above the polymers' softening point in the presence of the inert gas, then *in situ* polymerization is comprised of *in situ* monomer and nano clay interaction, and polymerization is comprised of initiation, propagation, and termination, then there are three methods namely: surface-initiated controlled/living radical polymerization (SI-CLRP), controlled radical-mediated photopolymerization (P-CRP), click coupling chemistry, and miniemulsion polymerization [12].

In Nigeria, the nano clays were mixed by hydration with various clay soils from Dogon-ruwa, functionalized, and characterized by the use of Raman spectroscopy, thermogravimetry, Brunauer Emmett Teller (BET), and particle sizer [13], then they were used as a partial replacement of cement in the road concrete pavement construction. Furthermore, the nano clay was dehydroxylated at 720°C, then characterized, and XRF equipment was used for the particle geometry; thus, it showed that the concrete formed was efficient under stormwater control and was recommended for low axle or low trafficked road design and construction, as well as aquifer recharge based on the flexural strength [14]. In Algeria, it was confirmed that when the clay soils were synthesized by ultrasonication with poly-glycidyl methacrylate (GMA) or nano clay composites, thermal stabilities were improved [15, 16].

1.2.3 Soil stabilization

This refers to making soil more robust and waterproof for a particular purpose, and there are two types of stabilization, namely surface (when the influence zone is less than 1 mm) and ground improvement (when the influence zone is more than 1 mm).

Surface stabilization consists of mechanical, physical, chemical, and physiochemical. Mechanical stabilization is where mechanical energy is used, for example, rollers, plate compactors, and tampers, and physical is done by cement, lime, bitumen, chemicals, and resin. Chemical stabilization is done by adding chemicals, for example, calcium chloride, sodium chloride, sodium silicate, polymers, and chrome lignin, and ground improvement consists of deep compaction, soil replacement, preloading, draining and GWT control, injection grouting, soil freezing, and use of geotextiles. Deep may be done by dynamic compaction or consolidation, vibro-compaction, or compaction piles, and soil replacement is when a weak/soft/organic soil is removed and replaced with compacted engineering soil. Preloading is done when 1.2–1.3 times the designed load is applied to allow desired settlement and accelerate consolidation. Drainage and GWT are where drains attain groundwater conditions, and lowering GWT or sometimes blanket drains and vertical sand drains are employed to accelerate the consolidation process. Injection grouting is where various fluid grouts are injected into the boreholes/weak soils and replaced by special pressure techniques, then soil freezing geotextile is correctly embedded in the soil and contributes to its stability.

In ancient times, lime was first used around 3000 B.C. for the Shensi Pyramids in China and 6500 B.C. in Syria [17–19], then in the seventh century B.C., the Chinese used lime in the construction of the Great Wall, bridges, underground chambers, and the clay or gravel soils were stabilized by the use of lime [5, 7, 20, 21]. In 1920, in the USA and Germany, it is where soil stabilization began in modern times, but the first tests on stabilization were done in the USA in 1904 [8, 17, 22, 23]. So, in order to demonstrate the solution in Africa, stabilization research has been done in the following countries.

1.2.3.1 Sandy soils

In the western provinces of Zambia (Mongu, Senanga, and Sesheke) in order to demonstrate the solution to the scarcity of road building materials, the sand samples were characterized, stabilized with cement and bitumen according to the design mixes selected, and California bearing ratio (CBR), unconfined compressive strength (UCS), indirect tensile strength (ITS), and Marshall stability tests were conducted, thus the sand properties were improved [24]. Additionally, in the Southern Sahara desert of Africa, during stabilization mixtures of high erodible soils, bentonite and, kaolinite were wetted with simulated rain, dried, and tested with wind tunnel with a brader at a free stream wind speed of 14 m/s or without abrader (wind only), and the results showed that the crushing resistance increased significantly with increasing clay content when a brader is available, then at the same time soil loss increased whenever the clay content increased, and furthermore, the loss ranged from 20 to 30 times more than the one from the trays treated with bentonite per kg of sand. Bentonite was more effective than kaolinite, and wind susceptibility was greatly reduced.

In Uganda, due to the sand abundance locally, clay samples were mixed with sand % proportions between 20 and 80%, and numerous laboratory experiments were conducted. The results confirmed that sand blending diminishes the shrinkage behavior of clayey soils. The properties of the clayey soils such as shrinkage behavior diminished, plasticity index and shrinking potential decreased, MDD increased, OMC decreased, unconfined compressive strength decreased, internal friction angle concerning shear strength parameters was enhanced, soil cohesion decreased, and consolidation settlement was lowered [25].

1.2.3.2 Restoration

In Ksar of Ait Benhadou, Morocco, the samples from the structure, were collected and analyzed by X-diffraction and X-ray fluorescence, then a representative sample was stabilized with three aggregates (lime, cement, and straw), and minerals, such as calcite and quartz, were encountered. They were rich in iron and potassium content and had a low plasticity (P.I. = 7%), which is slightly lower than the plasticity value required by the Moroccan standard for earth constructions. Generally, the results showed that the stabilized clayey soil properties were improved such as density, porosity, water absorption, and high thermal insulation [26].

In Namibia, the study of tailings left un-rehabilitated after 200 mines were exploited, considered the successful stories and lessons from other phytoremediation work, specifically phyto stabilization projects, which dealt with restoring, contaminated mine tailings. The study analyzed the OMT results using X-ray fluorescence geochemical data to understand the heavy metals in the tailings, and it further used GIS to assess the distribution of heavy metals on OMT, specifically those with high phytotoxic levels. Finally, the study addressed its primary objective, which was to provide the Namibian government with suggestions and recommendations on the best remedial measures for the remediation of heavy metal-contaminated mine tailings [19].

1.2.3.3 Heaving expansive clays

In Egypt, representative samples of 10% of GGBS and the replacement of 30% of hydrated lime were mixed with clay content and then cured within representative conditions: 20°C with 90–100% relative humidity, 350°C with 50–60% relative humidity for 12 months, after which compaction, swelling, plasticity, and UCS tested were conducted. X-ray diffraction, scanning electron microscopy, differential thermal analysis, and nuclear magnetic resonance (NMRC) were used to determine the reaction products. Results showed that engineering properties improved; the addition of lime showed a more significant improvement, maximum dry density (MDD) decreased, and optimum moisture content (OMC) increased with increasing GGBS [27]. Then in Morocco (Kenitra, Sidi Kacem, Tangier, and Tetouan) clay samples containing various minerals, such as hydrated Lime, montmorillonite, illite, and kaolinite, were mixed with pure standard clays, and a series of laboratory tests were conducted such as Atterberg limits, X-ray diffraction, hydration heats, and cementing; hence, there was a reduction in plasticity and an improvement in compaction properties, and the amount of lime needed to modify clay soil varied from 3 to 6% [28].

In Nakapiripirit district, (Northern Uganda), expansive soil samples were mixed with EAF slag dosages of 0, 5, 10, 15, 20, and 30%, after which laboratory tests were conducted on each mixture, and the expansive clay properties were improved (**Table 1**) [48].

No	Country	Problem	Materials, procedure, results, and conclusions
1.0	Ghana	Weak sub-base soils	*Materials* • Laterite • Lime • Cement *Procedure* • A refilling box was used to divide the sample into two equal parts, whereby one part was again divided into two equal parts, and this was done continuously until a proper level of mixing had been attained. *Results* • A total of 6% addition of lime to the sample resulted in P.I., L.L., and CBR values that passed GHA specifications for both base and sub-base courses [29].
2.0	Kwali Area Council in Abuja, Nigeria	Effect of stabilizing lateritic soil with a combination of bitumen emulsion and cement.	*Materials* • Lateritic soils • Bitumen emulsion • Cement *Procedure* • The additives (4, 6, and 8%), bitumen emulsion, and cement (100:0, 75:25, 50:50, 25:75, and 0:100) were considered, and UCS and CBR tests were determined. *Results* • Both values increased as the cement component increased for both soil samples, and so the strength of the soil improved [30].
3.0	Nigeria	Geotechnical properties	*Materials* • Periwinkle shell powder (PSP) • Lateritic soil. *Procedure* • 2, 4, 6, 8, and 10% of PSP and OPC were added to the un-stabilized lateritic soils, and NMC, P.L., and P.I. tests were conducted, respectively. *Results* • An increase in maximum dry density (MDD) and OMC was observed whenever cement and periwinkle shell powder (PSP) were added, and PSP recorded an increase of 5.6% in CBR value compared with OPC, which recorded a rise of 34% in CBR value, so periwinkle shell powder (PSP) can be used as a good stabilizer for clayey or lateritic [31].

<table>
<tr><th>No</th><th>Country</th><th>Problem</th><th>Materials, procedure, results, and conclusions</th></tr>
<tr><td>4.0</td><td>Kumasi Ghana</td><td>Heavy metal pollution</td><td>Materials
• Low-grade CaO
• Heavy metals
• Leachate
Procedure
• Five mixtures of leachate (each 100 ml) and LG-CaO (10, 15, 20, 25, and 30 g) were prepared, and weighed and mixing was carried out with magnetic stirrers at 110 rpm for 8 hours, then the samples were kept at 21o°C for 21 days [20, 32, 33].
• Soil samples each weighing 100 gl were mixed with a % LG-CaO such as 10, 15, 20, 25, and 30%, and each mixture was stirred by use of magnetic stirrers at 110 rpm for eight hours, then the mixtures were kept at 2100°C for 21 days [22, 34].
• Then 10 ml and 100 g of distilled water were used to mix each mixture to remove minerals that may have been caused by the tap water [35].
• In the second part, 100 g and 10 g of sand, LG-CaO contents were kept constant by weight while the water proportions [2, 27, 30, 36, 37] varied. So, after blending, compaction was done, and samples were kept at 250°C for 21 days [20, 32, 33]. After the heavy metal stabilization, the mixtures were again digested by use of triacid mixture. The concentrations of metals (Cd, Fe, Zn, and Cu) were analyzed by using an atomic absorption spectrometer (SPECTRA AA 220 Air-acetylene Flame) [38, 39].
• After 1, 5, 10, 17, and 21 days, the electrical conductivity, pH, and temperature of the mixtures were always measured before and after the stabilization treatment with a palintest multipurpose pH meter [11, 40, 41].
Results
• Concentrations of heavy metals (Cd, Cu, Fe and Zn) in the soil investigated were found to be above the environmental protection agency (EPA) and World Health Organization (WHO) guidelines, which can be a potential source for some heavy metals in the environment. In contrast, the physicochemical properties (temperature, clay content, moisture content, electrical conductivity, and pH) of soil samples and the leachate sampled were below the EPA threshold values. An increase in LG-CaO to the sample while maintaining the sand and water proportions constant to increase the P.H. or varying water proportion keeping sand and CaO constant led to a reduction of the heavy metal concentration within the soils after stabilization treatment [42].</td></tr>
</table>

No	Country	Problem	Materials, procedure, results, and conclusions
5.0	Africa	Cost-effective stabilization	*Materials* • Soil Kilned • Powdered Glass Wastes *Procedure* • The soil samples were exposed to an air room at room temperature and dried thoroughly, then they were disintegrated, and representative samples were selected to be used for the required tests. After this, the samples were kilned at a temperature of 700°C for 2 hours then cooled for 24 hours for further tests. A #200-micron sieve was used for sieving after having crushed the soil. • The % proportions of the mixture were 75% soil kilned and 25% powdered glass waste (3:1). Then % of the stabilizers varied, for example, 5, 15, and 25% of the total weight of the sample. *Results* • MDD, CBR, and UCS values were increased, while the OMC decreased after the 14 days of curing. Additionally, a free swell of expansive soil and CBR swell values were improved, and montmorillonite illite minerals were used for the X-ray di¢raction (XRD) test to have disappeared. Montmorillonite and illite minerals disappeared [43].
6.0	North of Morocco	Sliding phenomena	*Materials* • Reinforced concrete piles • Finite element modeling *Procedure* • Finite element modeling was used to check reinforced concrete piles, and the elastoplastic Mohr-Columb model was used for soil modeling. Inclinometers were installed on the site to measure soil deformation over time. *Results* • The horizontal displacement of the soil was close to the measured *in situ*. Also, the measured displacements revealed that the method of reinforcement by piles used in some areas of the study effectively stabilizes landslides, and not others [44].

No	Country	Problem	Materials, procedure, results, and conclusions
8.0	All countries	Environmentally friendly and sustainable stabilizers.	*Materials* • Waste tires *Procedure* • Clays mixed with an optimum of the waste tire 20% by weight gave better performance results than those from Geogrid. • Additionally, engineering properties were improved after adding 2% waste tire rubber fibers to cement-stabilized bentonite clays. Then, UCS ductility behavior was also achieved. • (30–50%) content coarse (4.75–2.00 mm) shredded tire waste improved the expansive black cotton soil geotechnical properties than the fine shredded tire, and it was observed that soil reinforced with a shredded tire tested by use of a square model footing showed that the bearing capacity of shredded tire reinforced soil increased by 2.68 times when 5% of soil-shredded tire mixture was used but when 5% of the soil-shredded tire mixture is added, there was a reduction in the bearing capacity caused due to excess shredded tire creating voids leading to the settlement of the foundations. • During stabilization mix ratios of clay soils named kaoline with tire crumbles, fly ash tire was mixed, and after mixing, the MDD decreased and OMC increased leading to a maximum bearing capacity which ranged from 5 to 20%. Furthermore, clay was mixed with sand soils and then stabilized by use of waste tire textile of % ranging from 0.5 to 4% thus leading to an increased bearing capacity [45].
9.0	Keruing sawdust	Environmentally friendly and sustainable stabilizers.	*Materials* • Saw dust *Procedure* • A total of 3% of Keruing sawdust mixed with expansive soils improved the expansive soil's properties, but durability tests were recommended. • A total of 0, 2.5, 5, 7.5, 10% of sawdust and 12.5% of the dry unit weight soil were used to stabilize the expansive soil, and the swelling potential and swelling pressure decreased with the increased percentages of sawdust addition. UCS increased sawdust addition by 7.5%, and beyond that, they started reducing. • Furthermore, different quantities of sawdust (1, 2, 3, and 5%) were also used to stabilize the expansive soils, and it was found that smaller percentages of sawdust brought a tremendous increase in the strength of the stabilized soils and 3% of optimum portion of sawdust was influential in the stabilization of the expansive soil, then beyond that there was a decline in strength observed, so it was concluded that sawdust could be used to fill the voids in soils. • In Southwestern Nigeria, various saw dust ash %, such as 0, 2, 4, 6, and 8%, were used to stabilize the soils, and optimal results were obtained when 6% of SDA was used. Additionally, 4% of the dry weight of the soil was used to stabilize the soil, which improved the properties. So, the CBR in both soaked and unsoaked conditions improved when 70% lateritic soil and 30% saw mixture were formed, and when percentages from 0, 2, 4, 6, and 8% by the dry unit weight of soil were used, the standard hydraulic conductivity requirement was improved, but 8% was the optimum content [45].

No	Country	Problem	Materials, procedure, results, and conclusions
10.0		Environmentally friendly and sustainable stabilizers	*Materials* • Fly ash *Procedure* • The collapsibility potential of gypseous soil decreased when stabilized by the fly ash class F activated with KOH and NaOH, and it was identified that geopolymer fly ash had higher sulfate resistance than Portland cement. • 0, 5, 10, 15, 20, and 30% fly ash were incorporated, and there was an increase in the maximum dry unit but a reduction in the optimum moisture content. *Results* • Whenever fly ash was increased and the curing period lengthened, the UCS increased, thus the highest UCS was obtained when 30% of flash was used after 90 curing days, then the strength properties of marine soils stabilized with cement were improved than those stabilized by fly ash in the short run (7–28 days). • The soil Atterberg test results improved by reducing the plasticity index from 20.2% (non-stabilized) to 13.1% when the soils were stabilized with 15% fly ash calcium fly ash (0, 3, 6, 9, 12, and 15% of the dry unit weight), which was class C fly ash because the sum of SiO2, Al_2O_3, and Fe2O3 is 27.53, which is below 70% and palm oil fuel ash [45].
11.0	Lake Chad basin, Central Africa	Ceramics applications	*Materials* • Clays samples *Procedure* • Particle size distribution, specific surface area, Atterberg limits, swelling rate, and clay activities were conducted then also XRF and XRD techniques were also conducted. • Samples were fired in the range from 750 to 1250°C. • Color, sound test, weight loss, bulk density, firing shrinkage, and compression strength measured the firing characteristics. • Water absorption was determined during 3 months in water immersion. *Results* • Plasticity was high and made up of smectites associated with kaolinite, illites, quartz and feldspar, and a high proportion of silica and iron, but less alumina, high iron content responsible for the reddish color in firing. • So, from the results, bulk density, linear firing shrinkage and compressive strength, and water absorption were < 20% obtained, which showed that the clays were suitable for producing ceramics and tiles [46].

No	Country	Problem	Materials, procedure, results, and conclusions
12.0	Algeria, Morocco and Tunisia	potential use for ceramic	*Materials* • Clays soils *Results* • Illite ranged from 12 to 38%, and kaolinite was 12–17%, as seen from the three countries. • Tunisia was comprised of 15% smectite, 4% palygorskite, 30% quartz, and 15% calcite, but the other countries were not mentioned, so it means they had very little maybe. • Large amounts of iron (>5.6%), earth-alkaline oxides (>6.9%), and high values of LOI (>12%). • The plasticity (P.I.) ranged from 16 to 40%, and Algeria had the highest requiring particular attention and careful temperature control during drying to avoid the deformation and the formation of cracks in the ceramic bodies. In contrast, the Tunisian and Moroccan clays (P.I. = 18 and 16%, respectively) show acceptable behavior in shaping and drying. • Indeed, the amount of fraction upper 63 μm was less than 2%. The abundance of components, such as Fe_2O_3, CaO, MgO, K_2O and Na_2O, influences the main transformations during firing. *Conclusion* • The firing shrinkage, water absorption, and flexural strength were within the ceramic international standards (ISO) [47].

Table 1.
Other construction techniques related to clay soils in Africa.

1.3 Clinker

These are solids which appear as nodules or lumps measuring between 3 mm and 25 mm or 0.12–0.98 inches in diameter produced during the kilning stages when limestone is heated with clay within a temperature from 1400 to 1500°C, after which gypsum will be added and then grounded to obtain cement. This is also a binder in various cement products, namely ground granulated blast furnace slag, Pozzolana, silica fume, and composite, and the clinker components are an alite, belite, aluminate, then the types are sulfate resistant, low heat, white, low alkali, belite calcium-sulfo aluminate ternesite, etc.

In Malawi, Cote d'Ivoire, Cameroon, Algeria and Ghana, gas suspension systems were created. The clay calciner system was expected to substitute between 30 and 40% of the clinker in the final product and these products improved the cement quality and controlled carbon dioxide emissions by up to 40% tone. In Nigeria, Level 1 reduction in greenhouse gas emissions was achieved when the calcined clays, limestone powder and ordinary Portland cement were used, and additionally, and this solved the housing deficit around Africa [49].

1.4 Adsorbent

This is the accumulation of large molecular species at the surface of liquid or solid phase compared to the bulk, and it arises due to unbalanced or residual forces at the surface of the liquid or solid phase. Adsorption happens at the surface of a substance, and absorption means uniform distribution of the substance throughout the bulk, for example, low energy electron diffraction (LEED), photoelectron spectroscopy (PES), and scanning tunneling microscopy (STM). Adsorption can occur in liquids or solids, and there are two types of adsorptions, namely physical or physisorption and chemical or chemisorption, then adsorption processes are studied through isotherm, GIBBs, Freundlich isotherm, Langmuir isotherm, multilayer, and BET. The method of adsorption is useful both in industrial and domestic, for example, heterogeneous catalysis, removal of coloring material, ion exchange resins, adsorption indicators, gas masks, dyeing of cloth and *de humidizers*, etc.

In Algeria, activated clay was used as an adsorbent to remove methyl orange (MO) from an aqueous solution, and pseudo-second order and second order were used, but the kinetic process followed the pseudo-second-order model. Additionally, the Langmuir and Freundlich isotherms models were used in the data description, and the results showed that all models obtained a good correlation. So, with nearly 30 minutes of contact time, the adsorption reached equilibrium and then became favorable at a lower pH [50].

2. Conclusion

The most significant findings are improvements in machine systems, procedures, and clay soil properties, and these will help to mitigate climate change during the manufacture of the products and execution of structures leading to saving of trees. However, although improvements were made to construction techniques related to clay soils, there is a need to implement them within the daily construction rather than only demonstrating most of them in the laboratory.

Author details

David Mubiru* and Sam Bulolo
Department of Civil and Environmental Engineering, Kyambogo University, Uganda

*Address all correspondence to: genius2gouganda@gmail.com

References

[1] Das BM, Sobhan K. In: Das BM, Sobhan K, editors. Principles of Geotechnical Engineering. SI ed. Stamford, CT, USA: Cengage Learning; 2014. p. 726

[2] British Standards Institution. BS 1377-2. Methods of Test for Soils for Civil Engineering Purposes—Part 2: Classification Tests. London: British Standards Institution; 1990

[3] Institution BS. N.A. to BS EN 1997-2 : 2007 U.K. National Annex to Eurocode 7: Geotechnical Design—Part 2: Ground Investigation and Testing. London: Bsi; 2007. pp. 1-11

[4] BSI. Methods of Test for Soils for Civil Engineering Purposes -Shear Strenght Tests (Total). London: BSI-1377-7; 2022

[5] BSI 1377-5. Methods of Test for Soils for Civil Engineering Purposes-Compressibility, Permeability and Durability Tests. London: British Standards Institution; 1990

[6] Huat BBK. Deformation and shear strength characteristics of some tropical peat and organic soils. Pertanika Journal of Science & Technology. 2006;**14**(2):61-74

[7] Twesigye-omwe MN, Kaweesa P, Kyakula M. A study of the use of ground investigation reports during foundation design in Kampala. In: Second International Conference on Advances in Engineering and Technology held at Entebbe, Uganda, January 31-February 1, 2011. Kampala: MacMillan; 2009. pp. 443-449. ISBN 978-9970-214-00-7

[8] Houben H, Guillaud H. Earth Construction—A Comprehensive Guide. London: Intermediate Technology Publications; 1994

[9] Easton D. The Rammed Earth House. White River Junction, Vermont, USA: Chelsea Publishing Company; 1996

[10] Keable J. Rammed Earth Standards ODA FINAL PROJECT REPORT 4864C; 1994

[11] Pérez-Peña AM, United Nations Human Settlements Programme and Good Earth Trust. Interlocking Stabilized Soil Blocks: Appropriate Earth Technologies in Uganda. Nairobi, Kenya: United Nations Human Settlements Programme (UN-HABITAT); 2009. Available from: infohabitat@unhabitat.org, www.unhabitat.org

[12] Guo F, Aryana S, Han Y, Jiao Y. A review of the synthesis and applications of polymer-nanoclay composites. Applied Sciences (Switzerland). 2018;**8**(9):1-29. DOI: 10.3390/app8091696

[13] Elele UU, Yakubu A, Ezekiel B, Gimba A. Synthesis, Characterization and In Vitro Antibacterial Activities of Some Amino Acids Functionalized Nanoclays from Two Local Deposits. 2020;**V**(Xii):147-158

[14] Ayobami B, Julius N, Emmanuel S, Snyman J, Williams K, Olugbenga A, et al. Permeable pavements for storm water control incorporating nano clay. Cogent Engineering. 2021;**8**(1):1-10. DOI: 10.1080/23311916.2020.1817254

[15] Cherifi Z, Boukoussa B, Zaoui A, Belbachir M, Meghabar R. Structural, morphological and thermal properties of nanocomposites poly (GMA)/clay prepared by ultrasound and in-situ polymerization.

Ultrasonics Sonochemistry. 2018;**48**:188-198

[16] Rane AV, Kanny K, Abitha VK, Patil SS, Thomas S. Clay–polymer composites: Design of Clay Polymer Nanocomposite by mixing. In: Clay-Polymer Nanocomposites. Amsterdam: Elsevier; 2017. pp. 113-144

[17] Bundesverband des Deutschen Kalkindustrie. Bodenverbesserung Bodenverfestigung mit Kalk. Köln: kalk innovative seit jahrtausenden; 2013. Available from: https://www.kalk.de/wissensportal/publikationen/fachpublikationen/erd-und-strassenbau/bodenbehandlung-mit-kalk

[18] Greaves HM. An introduction to lime stabilization. In: Rogers CDF, Glendinning S, Dixon N, editors. Proceedings of the Seminar. Loughborough, UK: Transportation Research Board, The National Academies of Sciences, Engineering, and Medicine; 1996. pp. 5-12

[19] Mbingeneeko F. Recommended Guidelines to Remediate Heavy Metals Contaminated Mine Tailings in Namibia. Namibia: Geological Survey of Namibia, Ministry of Mines and Energy; 2014

[20] Green VS, Stott DE, Diack M. Assay for fluorescein diacetate hydrolytic activity: Optimization for soil samples. Soil Biology and Biochemistry. 2006;**38**(4):693-701

[21] Van den Kerkhof E, Van Den Bergh H, Verhasselt A. Amélioration des sols à la chaux. Bulletin CRR. 2001;**47**:3-7

[22] Bürger W. Verfestigung mit Kalk in bindigen Böden der Bundesrepublik. Strassenbau-Technik. Ausgabe. 1972;**25**(11):27-34

[23] Johnson AM. Use of Lime in Improving Bases and Subgrades. Engineering Bulletin Purdue University: Extension Series No 69. USA: Purdue University; n.d. pp. 77-92

[24] Chilukwa NN, Hendrix W. Stabilization of Western Province Sands. 2016;**7**(8):247-253

[25] Jjuuko S, Kalumba D, Bagampadde U. The Use of Locally Available Sand in Stabilization of Ugandan Clayey Soils: Case Study of Clayey Soil from Busega Area. Academia Edu; 2011. Available from: https://nru.uncst.go.ug/handle/123456789/7856, uncstresearch@uncst.go.ug

[26] Lechheb M, Haarou A, El Boukili G, Azrour M, Lahmar A, El Ouahabi M, et al. Physico-chemical, mineralogical, and technological charactrisation of stabilized clay bricks for restoration of Kasbah Ait Benhadou-Ouarzazate (south-east of Morocco). Materials Today. 2022;**58**(4):1229-1234

[27] Mohamed El-Sadek Abdel Rahman Ouf. Stabilization of Clay Subgrade Soils Using Ground Granulated Blastfurnace Slag. Scanning Electron Microscopy; 2001. Available from: https://api.semanticscholar.org/CorpusID:135633906

[28] Maissara J, Karym H, Chbihi MEM, Belaaouad S, Moutaabbid M. Chemical Stabilization with Lime of Expansive Clays from Northern Morocco. 2016;**3**(January):1-6

[29] Achampong F, Anum RA, Boadu F, Djangmah NB, Chegbele LP. Chemical stabilization of laterite soils for road construction. Case study: The laterite soils at legion. International Journal of Scientific & Engineering Research. 2013;**4**(11):2019-2041

[30] Gbenga Matthew A, Shola Paul A. Soil stabilization using bitumen

emulsion and cement combination as additive. Journal of Earth Science and Engineering. 2018;**8**:66-74. DOI: 10.17265/2159-581X/2018.02.000

[31] Olowofoyeku A, Ofuyatan O, Oluwafemi J, Iroaganachi P. Geotechnical properties of lateritic soil stabilized with sugarcane straw ash. Journal of Physics: Conference Series. 2021;**1734**(1):2014-2025. DOI: 10.1088/1742-6596/1734/1/012011

[32] Spokas KA. Review of the stability of biochar in soils: Predictability of O: c molar ratios. Carbon Management. 2010;**1**(2):289-303

[33] Kolias S, Kasselouri-Rigopoulou V, Karaholics A. Stabilization of clayey soils with high calcium fly ash and cement. Cement and Concrete Composites. 2005;**27**(2):301-313

[34] Mahmoudi S, Bennour A. Characterization and ceramic application of clays from North Africa. Applied Earth Science: Transactions of the Institute of Mining and Metallurgy. 2022;**131**(1):15-26. DOI: 10.1080/25726838.2021.1992815

[35] Wang YM, Chen TC, Yeh KJ, Shue MF. Stabilization of an elevated heavy metal contaminated site. Journal of Hazardous Materials. 2001;**88**(1):153-159

[36] Jjuuko S, Kalumba D. South Sudan evaluation of index and strength characteristics of clay soil stabilized with lime as subgrade material in Jonglei, South Sudan. In: 16th African Regional Conference on Soil Mechanics and Geotechnical Engineering, Tunisia. 2015. DOI: 10.13140/RG.2.1.3808.2727

[37] McDowell C. Stabilization of soils with lime, lime-fly ash, and other lime reactive materials. In: McDowell C, editor. Lime and Lime-Fly Ash as Soil Stabilizers (Highway Research Board Bulletin No 231). Washington, DC: Highway Research Board; 1959. pp. 60-66

[38] Maurya A, Kesharwani L, Mishra MK. Analysis of heavy metal in soil through atomic absorption spectroscopy for forensic consideration. International Journal for Research in Applied Science and Engineering Technology (IJRASET). 2018;**6**(6):1188-1192

[39] Srikanth P, Somasekhar SA, Kanthi GK, Babu KR. Analysis of heavy metals by using atomic absorption spectroscopy from the samples taken around Visakhapatnam. International Journal of Environment, Ecology, Family and Urban Studies. 2013;**3**(1):127-132

[40] Jaime Graterol M. Scholars' mine histories in geotechnical engineering a review of foundation failures on plastic clays, following the yield shear strength concept of a plastic solid in this kind of soil. In: 6th International Conference on Case Histories in Geotechnical Engineering, Missouri University of Science and Technology, Arligton, VA. 2008

[41] Schofield RK, Taylor A. The measurement of soil pH. Soil Science Society of America. 1955;**2**(19):164-167

[42] Odom F, Gikunoo E, Arthur EK, Agyemang FO, Mensah-Darkwa K. Stabilization of heavy metals in soil and leachate at Dompoase landfill site in Ghana. Environmental Challenges. 2021;**5**:100308

[43] Tsegaye Woldesenbet T. Experimental study on stabilized expansive soil by blending parts of the soil kilned and powdered glass wastes. Advances in Civil Engineering. 2022;**18**: 1-12. DOI: 10.1155/2022/9645589

[44] Bouajaj A, Bahi L, Ouadif L, Ejjaouani H, Awa M. Study of the displacement of a soil stabilized by piles using finite elements method. ARPN Journal of Engineering and Applied Sciences. 2016;**11**(3):1793-1798

[45] Niyomukiza JB, Wardani SPR, Setiadji BH. Recent advances in the stabilization of expansive soils using waste materials: A review. IOP Conference Series: Earth and Environmental Science. 2021;**623**(1):1-8. DOI: 10.1088/1755-1315/623/1/012099

[46] Temga JP, Mache JR, Madi AB, Nguetnkamc JP, Bitom DL. Ceramics applications of clay in Lake Chad Basin, Central Africa. Applied Clay Science. 2019;**171**:118-132

[47] Bennour A, Mahmoudi S. Characterization and ceramic application of clays from North Africa. Applied Earth Science (Transactions of the Institutions of Mining and Metallurgy). 2022;**131**(1):15-26

[48] Nagasha J, Gitta D. Stabilization of Expansive Soils Using ELECTRIC-ARC Furnace Slag for Subgrade Construction. Margaret Trowell School of Industrial and Fine Arts (MTSIFA) Collection Makerere University. Uganda: Makerere University Institutional Repository Administrators; 2019. Available from: makud.mulib@mak.ac.ug

[49] Avatar, Akinyemi O. Revisiting Clay for Brick Making in Africa. mustardinsights; 2023. Available from: https://blog.mustardinsights.com/industries/revisiting-clay-for-brick-making-in-africa

[50] Bendaho D, Driss TA, Bassou D. Adsorption of acid dye onto activated Algerian clay. Bulletin of the Chemical Society of Ethiopia. 2017;**31**(1):51-62. DOI: 10.4314/bcse.v31i1.5